LEARNING
COMPUTER
GRAPHICS

Springer
New York
Berlin
Heidelberg
Barcelona
Budapest
Hong Kong
London
Milan
Paris
Singapore
Tokyo

Shalini Govil-Pai   Rajesh Pai

LEARNING COMPUTER GRAPHICS

From 3D Models to Animated Movies on Your PC

With 81 Illustrations

 Includes CD-ROM

 Springer

Shalini Govil-Pai (sgovil@hotmail.com)
Rajesh Pai (pai_raj@hotmail.com)
6309 Dana Street
Oakland, CA 94609
USA

Library of Congress Cataloging-in-Publication Data
Govil-Pai, Shalini.
    Learning computer graphics: from 3D models to animated movies on
your PC / Shalini Govil-Pai, Rajesh Pai.
        p.   cm.
    Includes bibliographical references (p.   )
    ISBN 0-387-94898-8 (softcover : alk. paper)
    1. Computer graphics.   I. Pai, Rajesh.   II. Title.
T385.G686   1998
006.6—dc21   98-24060

Printed on acid-free paper.

Production managed by Anthony K. Guardiola; manufacturing supervised by Jacqui Ashri.
Camera-ready copy prepared using the authors' FrameMaker files.
Printed and bound by Hamilton Printing Co., Rensselaer, NY.
Printed in the United States of America.

9 8 7 6 5 4 3 2 1

ISBN 0-387-94898-8 Springer-Verlag New York Berlin Heidelberg   SPIN 10556451

Preface

The field of computer graphics is rapidly expanding. With the increasing popularity and power of personal computers, computer graphics is growing at a phenomenal rate. Computer graphics applications are cropping up in diverse areas like entertainment, education, multimedia, and medicine, to name a few. The use of computer graphics effects in movies like *Jurassic Park* have dazzled millions of viewers. The success of such endeavors is prompting more and more people to use the medium of computer graphics to entertain, to educate, and to explore.

The growth of computer graphics (CG) is really a subset of the computer technology boom in the later part of this century. The spectacular growth and popularity of CG can be attributed to its applications being so "visual." The output from computer graphics as we witness it today in TV commercials, in movie special effects, in music videos, and in computer games, is just the tip of a huge iceberg of applications. CG is constantly evolving and attracting more and more attention. CG applications are multidisciplinary, blending together different regimes in art, science, and engineering. This allows for a unique and exciting merging of right brain logic and left brain creativity.

Just a decade ago, CG applications were restricted to high end workstations available to an elite few. Today with the rapid advances made in microprocessor speeds and other graphics hardware, personal computers can handle more and more complex graphics applications. The three-dimensional (3D) images output from CG applications are becoming increasingly photorealistic. Even high school students can work on their home PC to create dazzling computer graphics effects like morphing and animation.

To understand the basics of computer graphics, a beginner needs to get acquainted with fundamental concepts in

- *Modeling* — creating objects in three-dimensional space.
- *Animation* — assigning a time-varying geometry and behavior to the modeled object.
- *Rendering* — creating a photorealistic image of the modeled object.
- *Image Manipulation* — enhancing rendered images to produce desired special effects.

This book is organized to give the reader a clear and concise overview of the above basic principles in computer graphics. New concepts introduced in a chapter are illustrated by hands-on projects using the software provided. The chapters are organized as described below:

Chapter 1 provides an overview of computer graphics (CG) and how it has evolved. It includes an introduction to computer graphics terminology and definitions.

Chapter 2 describes what modeling means in CG. The concept of wire frame models is elucidated. Basic models (sphere, cube, cylinder, cone, polygon) are covered and an insight into polygonal representations of other complex objects is also provided. The projects included in this chapter involve use of modeling concepts learned in the chapter.

Chapter 3 discusses animation in detail. Principles of frame animation and real time animation are explained. The reader is given the opportunity to animate the modeled objects from Chapter 2.

Chapter 4 covers rendering of the wire frame objects created in Chapter 2. The fundamentals of lighting, shading, and texture mapping are discussed. The objects created in Chapter 2 are rendered by the user and the complete animation is seen in a rendered form.

Chapter 5 concentrates on creating special effects once images and animations are rendered. Basics of *compositing, warping, blending,* and *morphing* are discussed. Each of these post-production effects is demonstrated in the projects. The user is provided some digitized images with which to work. Interested reader can digitize and innovate using their own image files and create their own, cool special effects.

Chapter 6 includes a look into the future of CG applications and where this exciting field is heading. An overview of CG applications in various areas is provided.

The book includes several diagrams and pictures in order to make it easier for the reader to grasp new concepts. It also makes reading this book a fun experience. Interested readers can refer to the bibliography provided to explore more detailed depths of computer graphics.

The software accompanying this book is meant to give the reader a hands-on experience to grasp key fundamental concepts discussed in the chapters. The version of software included with this book requires an IBM-PC compatible with a 486, Pentium, or higher CPU, a hard drive, and at least 8 MB of RAM. The programs can produce graphical images on a VGA or SVGA monitor. An SVGA monitor is highly recommended for viewing images. The program will run on Windows 95 or higher operating systems. Support is provided for bitmap (.bmp) image files. Object models of either our proprietary mdl format or the DXF format put out by Adobe can be read.

This book can be used as a text (or course supplement) for a basic computer graphics course; the only prerequisite is that the students have access to a personal computer. The primary purpose of the book is to give students an overview of graphics and help them decide if they would like to pursue CG further, as a career.

The book can also serve well as an introductory book for interested professionals in different fields who would like to know more about the exciting field of computer graphics. The software projects will help them gain insight into CG for designing, developing, and experimenting.

In addition, this book can be an excellent reference for computer animators and artists who use computer graphics for their creations. This book will give them a broad overview of what is really going on behind the scenes in computer graphics software packages.

We hope you will enjoy reading the book and working with the software.

Acknowledgments

The inspiration to write this book stemmed from a number of people without whom the book would not be a success. After the popularity of movies like *The Lion King* and *Toy Story* several friends encouraged us to put together a comprehensive, hands-on computer graphics book. The book was a long and sometimes arduous journey. Our lovely daughter, Sonal, came into this world while writing this book. We would like to dedicate this book to Sonal and the several millions of kids who will be our future.

Special thanks to the highly skilled animators and software developers who provided inspiration and helped compile the software package for the book. We would also like to thank our editor Bill Sanders, the production editor, Tony Guardiola, and the friendly production staff at Springer-Verlag who helped us produce this book in its final form.

And finally, we would like to express our sincere gratitude and thanks to our parents, Mr. and Mrs. G. S. Pai, and Mr. and Mrs. G. Govil, who have always been our supporters and goaded us on to achieve our goals in life.

Berkeley, CA 1998   Shalini Govil-Pai
                                                                                                   Rajesh Pai

Contents

Preface   v

About the Authors   xi

1 Introduction   1
   1.1 The Basics   2
   1.2 Summary   8

2 Modeling   11
   2.1 Introduction   11
   2.2 Cartesian Coordinates   12
   2.3 Three Dimensional Modeling   17
   2.4 Primitive Shapes   21
   2.5 Drawing Objects on the Screen   30
   2.6 Transformations   35
   2.7 Hierarchy of Objects   42
   2.8 Advanced Modeling   46
   2.9 Summary   49

3 Animation   51
   3.1 Introduction   51
   3.2 Traditional Animation   52
   3.3 Concepts in 3D Computer Animation   53
   3.4 Animating Snowy, the Snowman   64
   3.5 Principles of Traditional Animation
        (In the Realm of Computer Animation)   66
   3.6 Viewpoint Animation   71
   3.7 Advanced Animation Techniques   76
   3.8 Summary   78

4 Rendering   **81**
  4.1 Introduction   81
  4.2 Hidden Surface Removal   82
  4.3 Lights   83
  4.4 Surface Materials   86
  4.5 Shading Algorithm   95
  4.6 Texture Mapping   98
  4.7 The Snowy Animation   103
  4.8 Summary   105

5 Postproduction   **107**
  5.1 Introduction   107
  5.2 The Images   108
  5.3 Image Enhancement   110
  5.4 Special Effects   118
  5.5 Summary   130

6 The Future   **131**

Appendix A   **135**
Appendix B   **139**
Appendix C   **145**
Bibliography   **147**
Index   **149**

About the Authors

Shalini Govil-Pai has received an MS in Computer Science from Penn State University and a Bachelor of Technology degree in Computer Science from the Indian Institute of Technology, Bombay. She has been working in the area of computer graphics and visualization for several years now. During this period she developed a wide range of graphics software including user interfaces, and visualization tools for Virtual Reality worlds. Currently she is working for PIXAR, a computer graphics and animation production house in Richmond, CA. She has worked on developing the technical aspects of commercials including the *Life Savers* and *Chips Ahoy* spots. She was involved in modeling objects and lighting scenes for the fully computer-animated Walt Disney feature film *Toy Story*. She is currently working on Pixar/Disney's second animated feature *A Bug's Life*.

Rajesh Pai has received his MS in Electrical Engineering from the University of Hawaii and a Bachelor of Technology degree in Electrical Engineering from the Indian Institute of Technology, Bombay. He also has an MBA from the Haas School of Business, University of California, Berkeley. Prior to business school, he worked as a senior program manager and design engineer at Intel Corporation for five years. At Intel, he was involved in designing and developing software solutions for the Pentium and Pentium II chips. He was also an active volunteer in Intel's community relations and corporate donations committee. In this role he worked in high schools to showcase emerging high technology products on personal computers to students and teachers. He currently works as an associate in a venture capital firm in San Francisco and evaluates software deals.

CHAPTER 1
Introduction

Computer Graphics: The term has become so widespread now that we rarely stop to think what it means. What is computer graphics? Simply defined, computer graphics is the graphic images created by a computer. Today's computers are capable of generating lifelike images virtually indistinguishable from images captured by photographs.

Everywhere you look these days you find computer graphics at work: from video games to movies to medical imaging systems in a hospital, to flight simulators used for military training. With the advent of the World Wide Web, everyone seems to be setting up Web sites displaying information about themselves or their company via text and graphics. All these applications depend on the basic underlying principles of computer graphics (CG, as it is often referred to) for their existence. A few decades back it was not possible to make computers draw on the screen. Only when monitors first started getting popular in the 1960s did computers acquire the ability to display images. Before that most of the computer input and output was through computer cards and printers. Now of course it is almost impossible to imagine using a computer without a monitor displaying the necessary information.

There are many ways to get images onto the computer. They can be hand drawn paintings or photographs that are scanned into the computer. Computer paint programs allow you to draw your images directly in the computer. You can also use the computer to define an imaginary world, with objects placed in this world. These objects can then be *rendered* by the computer to produce stunningly real looking images, like those seen in the movie *Toy Story*. A rendered image is built from the ground up starting with a mathematical description of the objects in the scene, to defining lights and colors to produce the final finished image.

Three-dimensional computer graphics deal with representing and rendering a three dimensional world that exists within the computer. Since the final result of a three dimensional computer graphics program is to produce a photograph like image, it borrows many ideas and principles from photography. When you click a photograph, you need a subject and you need lights to illuminate your subject. Finally you look through the camera viewfinder and click your photograph. Similarly, in CG you need to define a scene that you are going to photograph. You need one or more subjects called models. You need to define some surface material properties to these models to make them look good. You need to add lights to your scene, and finally you need to define a camera location and line of sight. In this computer simulated world, you are the artist with complete control over all the elements: the lights, models, and the camera. Based on your choice, the computer will generate the desired image for you.

In the next few chapters you will learn some of the technology involved in modeling three-dimensional objects, making them move, and giving them material properties. You will see how computers are used to generate three-dimensional scenes in a photorealistic manner. Before we jump into how three dimensional graphics work, it is important to form a foundation of the principles underlying how a computer stores information and how it displays simple figures and images. Once you understand these concepts, it will be easier to conceptualize what is happening inside the computer when we instruct it to draw our three-dimensional scenes. In the next section we talk about the basics of the computer and how it stores its information.

1.1 The Basics

Bits and Bytes

The computer stores numbers and values internally in a structure called a *bit*. A bit can have only two values, 0 or 1. This may not seem like much, but it is when bits are combined together to form larger values that one really appreciates their strength. Any number can be represented by a string of bits, each bit being assigned a value of 0 or 1. Each bit in the string is multiplied by an appropriate factor of 2. For example, the number 156 would be represented as shown in Fig. 1.1.

This is a string of 8 bits put together. A string of 8 bits has a special name in computer terminology. It is called a byte. Usually bytes is the standard for representing characters such as numbers or letters. Larger numbers may require two or more bytes to represent their value. Even

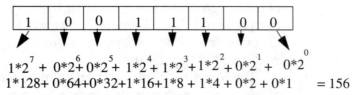

Fig. 1.1: The number 156 represented in bits.

floating- point numbers can be represented by 4 bytes, 2 for the abcissa (the part of the number before the decimal point), and 2 for the mantissa (the part of the number after the decimal point). You may have even seen that computer memory is depicted in terms of bytes. We say a computer has 16 Megabytes (MB) of RAM (Random access Memory) and a 600 MB hard drive. One MB is approximately a million bytes (it is actually 2^{20} = 1,048,576 bytes). This means that the computer has approximately 16 million bytes of RAM, and 600 million bytes of hard drive space.

The value stored in a byte can represent different kinds of information depending on the need of the user. A financial analyst may want to store information about the stock prices of a company; an astronomer may wish to store information about the trajectory of a comet. In computer graphics we are mainly concerned with two things:

- How does the computer store color information?
- How is this information organized to form images?

Let us start by exploring how we can make the computer represent color.

Color and Pixels

You may know that all colors can be represented by mixing in differing amounts of the three primary colors, red, green and blue. For example, yellow can be thought of as one part red, one part green, and no part of blue, orange can be a bit more of red than yellow and so on.

In this book, we employ a red, green, and blue (or RGB) color model. The RGB primaries are additive primaries; that is, the individual contributions of each primary are added together to yield the resultant color. We allow each primary to have a maximum value of 255 and a minimum of 0. This means that at a value of 0, we have no contribution from the color and at a value of 255, the color is at its brightest. We represent any color as a triplet (R,G,B). The color white contains equal contributions of the three primary colors at their maximum intensity. So in this RGB system, white would be represented as (255,255,255). Black has no contribution from any of the three colors, and is represented as (0,0,0). Yellow would be (255,255,0), which has maximum intensity of red and green and no contribution from blue. Of course this is yellow at its brightest. For a duller yellow we would simply reduce the R and G components. There exist other color models but the RGB model is well suited for our needs.

The computer displays information on the screen in what we call *pixels*. A pixel is the smallest variable element on a computer display (or in a computer image). The pixels are arranged on a display in rows and columns. A display screen has a certain maximum picture size that is defined by the boundaries of its physical size (e.g., a 13-inch monitor has a corner-to-opposite-corner distance of 13 inches), and the number of pixels specified horizontally and vertically (e.g., 640 pixels by 480 pixels), also known as the resolution of the display. The resolution of the display depends on the capability of the video card and the display monitor. The display screen is then actually composed of a grid of pixels.

A single pixel can be thought of as a box of light. It has three light variable elements, corresponding to the RGB color components, which can be varied in intensity to form the perceived color, as depicted in Fig.1.2.

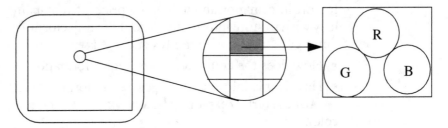

Fig. 1.2: The computer monitor and a component pixel.

Each pixel color can be set independently of the other pixels by varying its light elements. The number of colors that can be displayed on the screen at one time, however, is limited by the graphics hardware being used. A monochrome display allows only two colors to be displayed at one time, whereas a true color monitor is capable of displaying more than 16 million colors.

The maximum number of colors that can be displayed simultaneously is determined by the number of data bits set aside for each pixel in a region of memory called the *video buffer*. The number of bits set aside is also called the color resolution of the display. On true color systems, each pixel is represented by 24 bits of color information: 8 bits (or one byte) for red, 8 bits for green, and 8 bits for blue. Each byte can have a value from 0 to 255. The value of each byte in turn controls the intensity of the corresponding light element (either R, G, or B) of

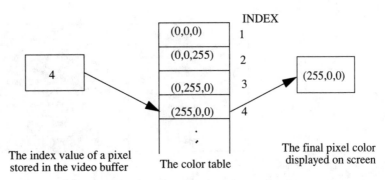

Fig. 1.3: How a pixel indexes into a color table.

the pixel. A 24-bit color resolution allows the system to display more than 16 million colors at once. Typically most VGA monitors are able to support only 256 colors at any given time.

On a 256 color system, only 8 bits are used in the video buffer for each pixel.This value is usually not the color value, but is an index into a color palette or color table. The color table is a table that stores the RGB components of colors in an indexed form. Any color can be accessed by determining its index position in the color table. In Fig. 1.3 we show some colors stored in a color table at different index positions. The index value stored in the video buffer is used to determine the color of the screen pixel. For example, in Fig. 1.3, the index value of 4 stored in the video buffer refers to color (255,0,0) in the color table.

The color table is independent of the video buffer and can be defined to have any number of bits to represent color information. If the table entries are defined to have 24 bits of color information, the index (an 8-bit value) can refer to a 24-bit color value. The video hardware can then select colors from a palette of 16 million colors, but it can display only a maximum of 256 colors at a given time.The advantage of using a color table is that it requires much less memory to make it work.

When a computer image is created or scanned in, the image stored is really a collection of pixels, as shown in Fig.1.4. Plate 1 shows the image and its zoomed in pixel view in color. The number of pixels horizontally and vertically determines the resolution of the image. The image has information about the color of each pixel in it. A true color image will have 24 bits of color information for each pixel, whereas a 256 color image will probably have 8 bits of information for each pixel which indexes into a color table of 256 colors defined for the image. The color table is specific to an image and is also referred to as the color map of the image. When displaying an image, the computer merely assigns the screen pixels to have the same color as that of the image pixels.

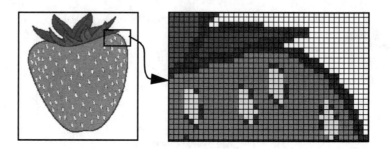

Fig. 1.4: A closeup of a sample bitmap reveals the pixels of the image.

In our examples we work with images which have a maximum of 256 colors. 256 colors is a good approximation for the total colors that the human eye can discern. In a later chapter we shall see how a true color image can be displayed on monitors having a lower color resolution.

Illuminating pixels sets them to different colors. How can this information be organized to form a meaningful figure like a line or a circle? In the next section we discuss a method used to organize pixels for this purpose.

Scan Converting Figures and Anti-Aliasing

The process by which a figure, like a line or a circle, is transformed into a set of data values for a group of pixels on the computer is called scan conversion. A variety of algorithms exist to scan convert basic geometric entities like the line. We shall not go into the math behind the algorithms. These algorithms tell the computer which pixels to turn *on* to draw the figure. For example, in Fig.1.5a we show a line passing through a grid of pixels. The pixels that the line passes through are turned on to represent the line, as shown in Fig. 1.5b.

One main problem in scan converting any figure is the jagged stair-like edge, as shown in Fig. 1.5c. This effect is known as aliasing. It is an unfortunate consequence of illuminating pixels either on or off. If the pixels are close together (a high resolution monitor), the staircase effect can barely be noticed. At a lower resolution, the harsh jaggies can lead to an unpleasant look.

A solution to this problem is called anti-aliasing. This solution involves illuminating pixels at varying intensities to blend in to form a smooth image. For example, in Fig.1.6a we show the same line again. This time the pixels are illuminated at different intensities depending on how much of the pixel is covered by the line, assuming the line is one pixel thick. The pixel intensity is represented by the darkness of

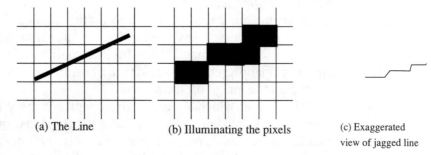

(a) The Line   (b) Illuminating the pixels   (c) Exaggerated
view of jagged line

Fig. 1.5: Scan Converting A line

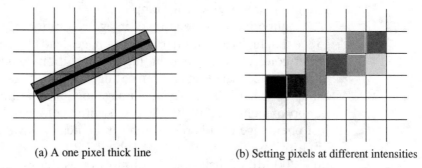

(a) A one pixel thick line   (b) Setting pixels at different intensities

Fig. 1.6: An Anti-aliasing solution.

each pixel, as shown in Fig. 1.6b. This approach leads to a blurrier but smoother image. The anti-aliased image does take a little longer to compute, as more pixels have to be turned on and off.

Once we have set the pixels to the colors and intensities to achieve the desired result, we would like to ultimately save these images in a file so we can refer to it again when needed. A variety of different formats to store computer images exist. Let us examine how image files are stored and what information is contained within.

Image Files

Transferring graphics images between different computers is routinely done in computer graphics. Most computer graphics people find the need to transfer images between different computers in order to use special programs to edit their images. The images must be imported and exported quickly and efficiently. To streamline this process, software developers have created a set of computer file formats that most graphics programs understand. Some of the common file formats are TIF, GIF, EPS, BMP, JPEG etc. Each of the different file formats have their own strengths and weaknesses, some of which we discuss in appendix A. Usually, we append the name of an image file with a tag to indicate its format. For example, a BMP image file called *myimage* would be saved as *myimage*.bmp.

Each image file needs to store information about the color resolution and size of the image. It should also store information about the color of each pixel in the image. With this information, it is easy to feed the file to a paint program that would then transmit the data to the computer to display the image on the screen.

The format which we use in this chapter is the Windows bitmap (bmp) file format developed by Microsoft. This file format is very popular on the personal computer and is supported by most programs that run on the Windows platform. It is also the default format for the Windows Paintbrush program, so you will not need any special paint program to paint or view your bmp images.

Coming back to the issue of image resolution, it may be obvious to you that high resolution true color images lead to crisper images with better color clarity. However, higher resolution comes at the cost of larger file size. Large images can be huge memory hogs, and can quickly fill up your hard disk. A typical image made for a film would need to be approximately 1,500 by 900 pixels with a color resolution of 24 bits. Just one image of this resolution can be nearly 1 MB in size. For a 5 minute film clip, you would need about 7,200 images. That would take up 7,200 MB of disk space! For most of us that would be enough to fill up our hard disks! Most production houses do employ compression techniques that can reduce the size of the image file. Depending on the type of compression used, and the complexity of the image, the image size can still be very large. Besides disk space requirements, a large file size will also typically take longer to load up into a graphics program and will also take longer to be written out to disk. In this book we normally work with images of about 320 by 200 pixels, with 256 colors, which is a fairly standard size. This is high enough resolution for our purposes and is not a disk hog compared to higher resolution images.

1.2 Summary

In this chapter we have covered most of the basic groundwork to understand the workings of the computer and how the computer stores and displays images. We saw how a computer used the RGB triplet value to depict the color at independent pixels, and how the pixel colors can be set to display color images. If you found yourself getting bogged down by the technical aspects, take heart. We provide the technical details for the sake of completeness. It is more important to understand the concepts behind how computer graphics work to appreciate the next few chapters.

We are now ready to explore the three dimensional world in a computer graphics setting and how to get the computer to generate our very own images!

CHAPTER 2
Modeling

2.1 Introduction

Computer graphics is built on three basic building blocks: modeling, animation, and rendering. Using these building blocks, we try and define a simulated three-dimensional world in the computer, which we can then *photograph* to yield desired images. The modeling phase involves defining the models or objects that reside in this world. We can then assign motion to these models (animation) and define them to have desired surface properties (rendering). Modeling, then, is usually the first step in the process of setting up our computer graphics world.

To model means to design, represent, or imitate an existing or proposed structure. You may be familiar with the model of our solar system. Fig. 2.1 shows a representation of the solar system model. We represent the sun as a sphere and the planets as smaller spheres orbiting around it.

A model represents features of a concrete or abstract entity. In computer graphics, a model is represented graphically. This graphical model helps us visually see the model's structure and observe the model's behavior, given certain dynamics of motion. When we model an object we tell the computer about the *shape* of the object in a three-dimensional world. Since computers work with numbers and mathematical formulae, objects that can be well defined with precise equations are easier to model. Some basic models are of shapes like a sphere, a cylinder, a cube, and a cone. These basic models have well defined equations to describe their shape and are commonly referred to as primitives. It is possible to create complex shapes by combining together these geometric primitives. Some models may need more advanced modeling

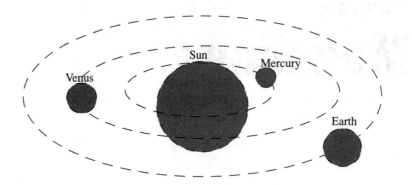

Fig. 2.1: A model of the Solar System.

techniques. We will discuss some of these methods at the end of the chapter.

How do we go about defining a simulated world with its models to the computer? Well, the field of computer graphics is not a totally original subject, for to define and resolve its problems, it draws upon well-established techniques from algebra, optics, geometry, and human physiology. Geometry is exploited to provide a framework for describing two-dimensional (2D) and three-dimensional (3D) space, while algebraic equations evaluate equations associated with this space. These equations are fed into the computer which then works on them to create your images. We will begin our discussion by looking into the basic framework used to represent the CG world and the models that exist in it.

2.2 Cartesian Coordinates

A key concept in CG is the definition of coordinate systems. The most common system used is called the cartesian or rectilinear coordinate system. In the two-dimensional case, the cartesian coordinate system enables a *point* on a flat surface to be addressed by using a pair of horizontal and vertical axes, which by convention are called the X and Y axes, respectively. The intersection point of these two axes is called the origin, and any point is located by measuring two distances parallel with the axes from the origin. The horizontal and vertical measurements for a point are called the x and y coordinates, respectively. Fig. 2.2 illustrates the scheme and shows the convention for positive and negative directions.

Using the coordinate system, any 2D shape can be represented by a sequence of points or *vertices* where each vertex is identified by a pair of (x,y) values called its coordinates. Fig. 2.3 shows a pentagonal figure with its vertices defined by points A, B, C, D, E in 2D space.

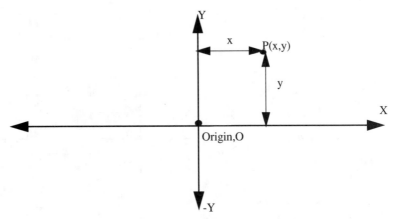

Fig. 2.2: The Two Dimensional Coordinate System.

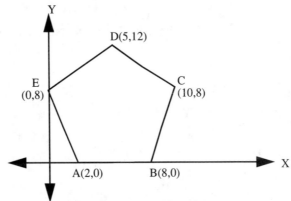

Fig. 2.3: The Coordinates of a Two-Dimensional Shape.

Let us extend the discussion of 2D space to three dimensions. We can easily relate to three-dimensional (3D) space because that is how we see the world. We know intuitively that every object has three dimensions: its length, height, and width. Extending cartesian coordinates into three dimensions requires that every point in three-dimensional space be located using three coordinate values, thereby defining all three dimensions of an object.

The three axes in a three-dimensional coordinate system are mutually perpendicular to each other (at 90 degrees with respect to one another). The point where they intersect is called the origin. The third axis is typically called the Z axis, as shown in Fig. 2.4.

In this text we use a righthanded coordinate system, as shown in Fig. 2.5a. Some texts follow the lefthanded coordinate system. The direction of the three axes in such a system is shown in Fig 2.5b. Throughout this book, a righthanded 3D system of axes is used for describing

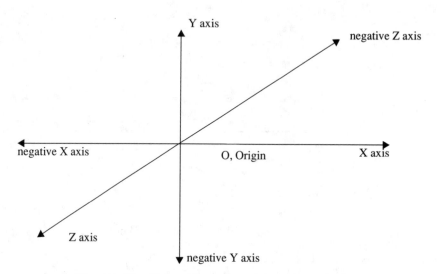

Fig. 2.4: The Three Dimensional Coordinate System.

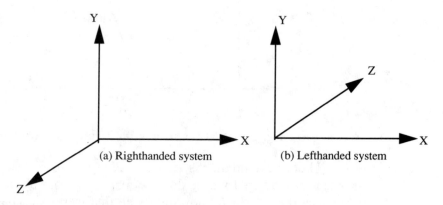

Fig. 2.5: The Left and Right Handed Systems.

the position of objects, cameras, and lights. In our discussions we will refer to this system as the world coordinate system (WCS).

At this point, let us start up the model software provided with this book. If you have not already done so, please refer to Appendix C on how to install the software onto your hard drive. Once you install the software, Windows95 users will see a new program group called 3dcg, under the Start->Programs menu. This group contains the icons to start all the software needed. WindowsNT users will see these icons in a new program group called 3dcg.

■ **Mini**
Project

- Start up the modeling software. For Windows95 users, do this by clicking on the model.exe option under the **Start->Programs->3dcg** menu. For Windows NT users, double-click the model.exe icon in the **3dcg** program group.

- You should see the form shown in Fig. 2.6.

- As we progress further in this chapter, you will learn how to exploit this form to create shapes and surfaces. The main window is called the Camera View window. This is where we shall see all the objects we are modeling. At the top is the command menu bar, that has various commands. Clicking on a command will bring up a menu of options associated with the command. You can select the desired option by pointing your mouse and clicking the mouse button

- Click on the **Show** menu and toggle the **Axes** option on. This will display the world coordinates in the camera view window for you. You should see the X,Y, and Z axes intersecting at the world origin. This is our world coordinate system.

- You can exit from the model program by clicking on the **File** menu and choosing the **Exit** option. Alternately, you can also exit the program by simultaneously holding down the **Ctrl** and the **D** button.

- Note all figures shown in this book for forms that appear using the software are for the Windows 95 platform. For some other operating system, the appearance of the forms may be slightly different.

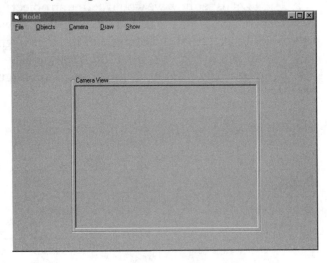

Fig. 2.6: The Modeling Interface.

One observation you may have already made is that some of the lines appear jagged. This is because we are not employing anti-aliasing techniques to smoothen these lines. We have defined anti-aliasing in the introduction chapter and you may want to refresh your understanding of this term.

In the world coordinate system, any point is located by three coordinates (x,y,z) representing the parallel distance along the three axes. This distance can be measured in feet, inches, or any other measure of distance. However, once we choose a particular unit we need to be consistent. In this book we will make use of the term *units* to refer to all measures of distance. Fig. 2.7 defines the x,y,z coordinates of a point P in 3D space.

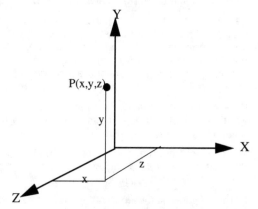

Fig. 2.7: A point represented in the 3D Coordinate System.

Let us see how we would define an object in our 3D space. Consider a box 5 units long, 3 units high, and 4 units wide. Let the corner of the box be placed at the origin and its sides aligned with the three axes, as shown in Fig. 2.8. Then the location (coordinates) of the vertices can be easily defined, as illustrated in the figure.

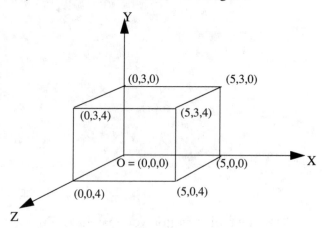

Fig. 2.8: The Vertices of a Box.

Each object has its own characteristic origin, about which it is typically said to be based. As shown in the above figure, the box has its origin at point O, which is coincident with the world origin. If we move the box around however, the local origin point of the box will no longer be coincident with the world origin. It is important not to get confused by the object's local origin, and the world origin and axes. The world origin and axes are always fixed. The local origin of the object stays fixed on the object, and hence moves with it. We shall look into this concept in greater detail when we discuss transformations.

Now that we understand the three dimensional space we are using to define our world, let us start dealing with the actual modeling.

2.3 Three Dimensional Modeling

Graphics scenes can contain different kinds of objects: trees, flowers, wood paneling, water, plastics, fire, etc. So it is probably not too surprising that there is no one standard method that we can use to describe objects that will encompass the characteristics of all these different kinds of objects. The most popular method to define models in space is to use polygons to approximate the surface of the modeled object. This method is popular due to the ease of use, speed of display, and the abundance of available algorithms to deal efficiently with the polygon-based models. In the next section we learn about polygons in more detail.

The Polygon

A polygon is like a cut out piece of cardboard. More formally, a polygon is a set of non-crossing straight lines joining co-planar (flat) points that enclose some single convex area. Let us examine this definition.

Drawing a polygon is like playing a game of connect the dots. The idea is to draw straight lines from one dot to the other in a specified sequence. You need at least three points to define a polygon. Lines making up the polygon should not cross. Each line of the polygon is called an edge, each corner a vertex.

The phrase *single convex area* refers to the plane that lies within the polygon. A single area means that the enclosed area should not be divided. The convex requirement means that given any two points within the polygon, you must be able to draw the straight line connecting those points, without going outside the area enclosed by the polygon, and always keeping your pen on the perimeter of the polygon. Fig. 2.9 shows some polygonal shapes generated by drawing straight lines con-

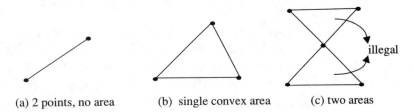

(a) 2 points, no area   (b) single convex area   (c) two areas

Fig. 2.9: Different shapes generated by connecting the dots.

necting the dots. Can you identify which shapes are legal polygons and which are not as per our definition? In general, the local origin of a polygon is said to be based at the center of the polygon.

To enter the polygon into the computer, you would type in the coordinates in exactly the order you want them joined to produce the polygon. A polygon can be multifaceted, i.e., have any number of vertices (greater than three). In this book we define all our polygons to have four vertices. Of course if you define two vertices to be equal, then the polygon defaults to a three-sided polygon. If you define more than two vertices to be equal, then the polygon is no longer a polygon. It will become a point or a line. When defining the polygon, you will have to be careful to ensure that the polygon is flat (i.e., all the vertices lie in one plane). The software will not check this for you.

A polygon has two faces, referred to as front face and back face. The order in which you specify the vertices of a polygon will define its front and back faces. The orientation of the vertices define a vector that is orthogonal to (that is, at 90 degrees with respect to) the plane of the polygon. This orthogonal vector is called the normal vector to the polygon, or the polygon normal. A vector is simply a line that has a direction, and is typically represented as a line with an arrow at its head, depicting its direction. The normal vector to the polygon points in the direction the polygon faces. If the vector points towards the camera, the polygon is called front facing; otherwise, it is called back facing. Figure 2.10 illustrates how the orientation of vertices defines the normal vector N, and hence the face of the polygon Some authors assume the same ori-

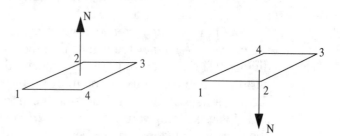

Fig. 2.10: The vertices of the polygon define the direction of its normal vector.

entation of vertices to define the polygon normal to be in the opposite direction. Both techniques are used widely and you will have to be aware of this when dealing with other software. A simple technique to identify the face of polygons in our software is to curl the fingers of your left hand in the direction of successive vertices defining the polygon. Your thumb would then point in the direction of the normal vector and hence identify the front face, and the heel of your hand points to the back face.

Note that three vertices are enough to define this direction. We shall look more deeply into the concept of normal vectors in a later section and also in later chapters.

Polygons have a limited range of usefulness by themselves. They are typically used to represent flat ground planes. However the real power of polygons comes when they are combined in a mesh structure to make a more complex model. A polygon mesh is a collection of polygons arranged so that each edge is shared by at most two polygons. A sheet of graph paper is a good example of a polygon mesh. The entire sheet would correspond to a modeled surface. Each square would correspond to a square polygon on that surface and each corner of each square is the vertex of the polygon. A polygon mesh is of course more flexible than this. The polygons need not be square, and they can exist in three dimensional space. The critical characteristic of a polygon mesh is that the polygons must be connected (i.e. share an edge with another polygon of the mesh). Consider our box. Each face can be thought of as a polygon. The box is then represented as a mesh of six polygons.

To model flat surfaces with polygons is easy. If you want the composite surface to be curved, (as shown for the cone in Fig. 2.11 things get a bit more tricky. Figure 2.11a shows the actual curved surface of the cone, and 2.11b shows the polygonal approximation for this curved

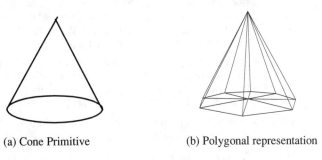

(a) Cone Primitive   (b) Polygonal representation

Fig. 2.11: A Cone Model and its polygonal representation.

surface. Since polygons are flat, you can never quite represent the curved surface exactly by using them. Using a larger number of polygons will approximate the curved surface more accurately but never exactly. Increasing the number of polygons can get very cumbersome as the model gets bigger, and will also take the computer longer to draw. There is a tradeoff between the accuracy of the model and the speed of display. Depending on the situation at hand, a compromise has to be settled for. There are tricks applied to handle bigger complex models, which we shall see in a later chapter when rendering the objects. The effect is that your eye sees a smoother surface than what is modeled.

Let us try some hands-on exercises that will reinforce the concepts we have studied so far.

Mini Project

- Restart the model program.

- For this exercise click on the **Objects** menu and select the **Create** option. You will be presented with a list of objects that you can create. Select the **Polygon** option. You will now be presented with a form, as shown in Fig. 2.12.

Fig. 2.12: Creating a polygon.

- The sliders will set the X, Y, and Z coordinates of the four vertices of the polygon. When the polygon is drawn initially, all its vertices are at the origin. So you may not see anything in the modeling window.

- Create a polygon with the following coordinates by sliding the X,Y,Z slider bars for each vertex of the polygon: Vertex1 = (15,0,0); Vertex2 = (45, 0,0); Vertex3 = (35,30,15); Vertex4 = (10,30,30). Note that as you edit all the points of the polygon, the points seem to grow out of the origin and as you add more points, these points will be joined by lines. After editing all the points of the polygon, you will be able to see the complete polygon in the modeling window. Click the OK button after you are happy with the shape of your polygon. Can you identify which face of the polygon you are looking at?

- You can try creating more polygons. While drawing polygons, remember that the order in which you define the vertices defines the orientation of the polygon. Although you can create nonconvex nonplanar polygons by sliding the coordinate sliders, it is recommended that you create only planar, convex polygons.

- Notice that we represent the polygon as a set of lines that connect its vertices. Such a representation is called a wire frame representation.

A *wire frame* representation is more formally defined as a collection of lines representing the major geometric features of the object. The wire frame model appears as though the object is physically constructed of straight pieces of wire. A wire frame model is a good place to start when modeling an object. The user can quickly see the shape and dimensions of the model being created. In the rendering chapter we will see how to define surface properties of a model to see a solid object.

Again, you may notice that some of the lines of the polygon look jagged, as no anti-aliasing techniques are being used. We will now proceed and define the primitive shapes using the versatile polygon shape.

2.4 Primitive Shapes

Basic geometric shapes that have a well-defined mathematical formula are typically referred to as primitive objects. The set of primitive objects normally includes a sphere, a cylinder, a box, a cone, and a disk. Although we use polygons to construct these shapes, they exist in their own right as primitive objects and most modeling software provides them as basic primitives. Throughout the book, we define just the surface of the model and not its interior volume.

Let us look at each of these shapes in more detail.

Box

The box model we have used so far in earlier examples is one of the most common shapes occurring in nature. In order to model a box we need to know its length (l), height (h), and width (w). Now we know that each object has its own local origin. The box has its local origin defined to be at its back left base vertex, as we saw in Fig. 2.8. If we define the local origin of the box to be the same as the world origin (O), then the location of its eight vertices can be uniquely defined by knowing the length, height, and width. The vertices of the box in turn define its polygonal faces.

■ **Mini Project**

- Assuming you still have the model program running, choose the Create option under the Objects menu and select the **Box** option. If you do not have model running, start it.

- You will be provided with a form to control the length, width, and height of this box object. The sliders let you set the length, height, and width of the box. Notice how the vertex points change as we slide these sliders around.

- Create a box with the dimensions length= 50; height = 25; width = 45.

- Can you formulate what the vertices of the box are, given the dimensions of the box?

- You must have noticed that when you created the box model, the polygon model you had earlier disappeared. Click on the Objects menu and if you select the **Pick** option, all the models you have created so far will be listed. Each model has a name associated with it (e.g., Polygon_0, Box_1). The number helps in assigning unique names to models with the same primitive shape (e.g., Polygon_0, Polygon_1). By clicking on a specific model, you can pick that model, causing it to be drawn on the screen. Pick the box object again.

- To modify the dimensions of a created model, click on the Objects menu and select the **Modify** option. This will bring up a form enabling you to modify the currently picked object. Pick the Modify option and try changing the length of your box to 60 units.

- The box's local origin is located at the world origin. Changing the length, height, or width will not change the position of this origin. Later on when we learn how to move objects around, you will see that the box's local origin moves with the box, while the world origin remains fixed.

Sphere

Baseballs, basketballs, and the earth are some common examples of spheres. Because of the symmetry of its shape, it is also one of the most simple objects to define in computer graphics. As shown in Fig. 2.13, in order to uniquely define a sphere we need to define its radius r, and

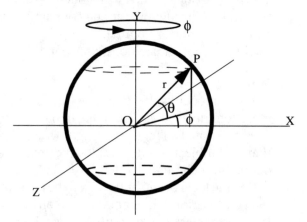

(a) Forming the Polygons on the surface of a sphere

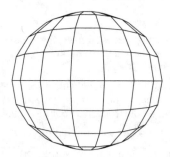

Fig. 2.13: Representing the Surface of a Sphere.

the location of its center (O). This center is the local origin of the sphere. We always define the sphere to have its local origin coincident with the world origin. This means that the sphere is centered around the world origin.

Any point P on the sphere can be described by the angles made by the line joining this point to the origin. We define the angle between the line and the XZ plane as θ (called theta), and the angle with the XY plane as ϕ (called phi). Refer to Fig. 2.13a. Then for a sphere with radius r, the coordinates of point P would be

$$P = (r * \cos \theta \cos \phi, \; r * \sin \theta, \; r \cos \theta * \sin \phi);$$

Note: Don't get overwhelmed by this equation; we just mention it here for completeness and will not be using it further.

Using the model explained above, we can go one full circle around the sphere if we advance angle ϕ from 0 to 360 in regular interval steps, keeping the angle θ fixed. At each interval point we can define a point $P(\theta, \phi)$. Now after completing one circle on the surface of the sphere, we can advance θ to a new value and then repeat the rotation of ϕ from 0 to 360 to define a new circle as shown in Fig. 2.13a. Now if θ were to vary from -90 to 90 at regular interval steps, we would cover the entire surface of the sphere, as shown in the figure. Each polygon gets described as a set of vertices calculated at different values of θ and ϕ. This means that if we were to connect all the points described, we would get a polygonal mesh representing the entire sphere (Fig. 2.13b).

Mini Project

- Go back to the model program. If you have stopped it from the last exercise, please restart it.

- Under the Objects menu, select the Create option and click on **Sphere**. A form comes up with a slider for the radius of the sphere. Slide the value of the radius up and down and see your sphere change its dimensions.

- Create a sphere of radius 25 units. Click on OK. Note that the local origin of the sphere is defined to be coincident with the world origin. Do not exit from the program as yet.

Before we discuss the other primitive objects, let us diverge for a moment to understand some important concepts in computer graphics using the sphere we just modeled as a visual aid.

Complexity

In the last exercise you may have noticed that the sphere displayed looks very faceted and is not smooth. Remember, we had talked about how we need more polygons to represent curved sufaces. The number of polygons that we define to represent the surface of an object is called the *complexity* of the object. Complexity hence defines the density of the polygon mesh. A higher complexity means a denser mesh of polygons and hence a more accurate model than one with a lower complexity. We continue with the mini-project.

Mini Project

- Under the **Draw** menu, click on the **Complexity** option. This will bring up a form with one slider, that controls the complexity of the objects being displayed. The default complexity is 1.

- Slide the scroll bar to different values. Notice how the sphere starts to look smoother and smoother as the number of polygons representing its surface increase in density. Fig. 2.14 shows a sphere drawn at two different complexity settings.

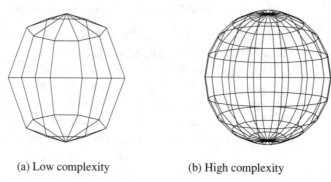

(a) Low complexity   (b) High complexity

Fig. 2.14: The sphere represented at different complexities.

Our complexity slider lets you set the complexity of objects to a number between 1 and 10. For different objects, this results in differing polygon density, the density increasing with higher complexity. You may notice that as you raise the complexity, the accuracy with which the sphere is represented increases. However, it also seems to take longer to redraw the sphere on the screen. At a complexity of 1, the sphere is drawn using about 32 polygons. At a complexity of 10, the sphere is drawn using nearly 1,000 polygons! The computer has to draw nearly 30 times the number of polygons when the complexity is raised from 1 to 10. Depending on the computer, it may not take the computer 30 times more time to draw this more complex sphere, but you can be sure it will definitely take longer to draw it on the screen. The vertices of the polygons are still drawn using the formula we provided. As we raise the complexity, the step interval we use between each successive θ and ϕ gets smaller, hence describing more polygons.

There is a tradeoff between performance and accuracy of a model. If you need very accurate models, you will take a hit in terms of performance. In several situations the accuracy of the model is not very crucial. All computer graphics professionals are concerned about the complexity-to-performance ratio. In realtime applications like computer games, speed is of vital interest to the graphics developers and they go through painstaking steps to reduce the polygon count of the models in their scene.

As we discussed earlier, most curved surfaces cannot be accurately represented by polygons. Such objects look more accurate as the complexity is raised. For flat surfaces, however, like the cube, increasing the polygon density has no effect on the accuracy of the model, as the initial representation was already an exact one. For these kinds of objects, we do not raise the polygon count as the complexity is raised.

Back-Face Culling

Another phenomenon you may notice while looking at your objects is that you can see all the polygonal faces of the model. You see the side of the model facing you as well as the side facing away from you. From real life we know that we cannot possibly see the side of a solid opaque object that is facing away from us. We need a technique to hide those polygons that' are facing away from the camera's viewpoint. This would help display a more realistic model of a solid object and also helps speed up the display time, as there are fewer polygons to be drawn at any given time.

We discussed the concept of polygons having a front face and a back face. The normal vector of the polygon, which points in the direction of the front face of the polygon, can determine on which side of the object the polygon lies. In Fig. 2.15 we show a horizontal slice of a low complexity sphere model. The figure shows a schematic of the

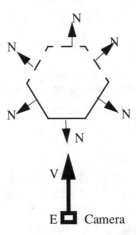

Fig. 2.15: A top view of The Camera and Polygonal faces of a sphere.

polygons and their normal vector directions. The camera is located at position E and is looking in the direction shown by V. To ensure the back face polygons get culled, we need to remove those polygons whose normal vector N point away from the eye of the camera. These are also the polygons whose back is facing the camera, shown in Fig. 2.15 by dotted lines. In computer graphics terminology this would mean that if the angle between V and the polygon normal N is less than 90 degrees (that is, the dot product of V and N is greater than zero), then the polygon is back-facing. This process of culling the back faces so we do not see the rear of the object, is called "back-face culling" (to cull means to remove). Note that our software defines the component polygons of primitive shapes in such a manner that their normals point in precisely the directions shown. Most computer graphics programs provide a way to describe polygons to be oriented in a way to implement back-face culling correctly.

- In the model program, with your sphere still being displayed, click on the Draw menu and select the **BackFace Culling** option. An arrow next to the option signifies that the option has been turned on. This enables back face culling, and you should now see the sphere as shown in Fig. 2.16b.

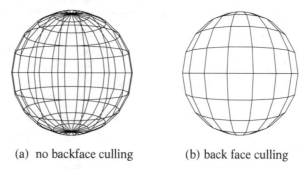

(a) no backface culling   (b) back face culling

Fig. 2.16: The sphere, with and without back face culling.

- Toggle back face culling off by clicking on it again. Notice how the back face polygons reappear. Keep the **BackFace Culling** option turned off for now.

Now that we understand these important concepts in modeling, let us continue with our discussion of the other primitive objects.

Disk

A disk is essentially a flat pizza pie; it has no height associated with it. Our disk is a 2D object in the XZ plane with Y = 0. A flat disk can be defined completely by its radius r and the location of its center (O), as shown in Fig. 2.17. We assume that all our disks have their local origin at their center, and this local origin coincides with the world origin. That is, the disks are drawn centered about the world origin. A wire mesh model of a disk is shown in Fig. 2.17. Any point on the perimeter of the disk can be defined by its angle θ from X axis. This point is defined as

$$P = (r * \cos(\theta), 0, r * \sin(\theta))$$

Note how in the above definition, the Y component for all polygon vertices is always zero, as the disk is flat and lies in the XZ plane. If we advance θ from 0 to 360 degrees by a fixed interval step, then we would traverse the perimeter of the disk, defining points on this perimeter. The polygonal mesh forming the disk is described by a set of points calculated at different values of θ. At higher values of complexity, like we saw in the sphere, θ between successive points is smaller defining more points and hence more polygons forming the disk. The center of the disk is located at the world origin by default.

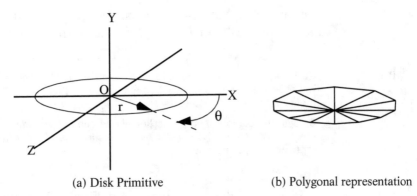

(a) Disk Primitive   (b) Polygonal representation

Fig. 2.17: The Disk model and its polygonal representaion.

**Mini
Project**

- Go back to the model program and under the Object menu select the Create option and choose the **Disk** option. Slide the radius slider of the form which pops up. Create a disk of radius 50 and click on OK.

- If you change the complexity of the object, notice how the angle between a slice of the disk changes.

Cylinder

A pencil, pen, the base of a tree trunk, and a telephone pole are all examples of the cylinder model. A cylinder can be defined by its radius (r), and height (h), as shown in Fig. 2.18. We assume that all our cylinders are located with the center of their base, which is also the local origin of the cylinder, located at the world origin (O) by default.

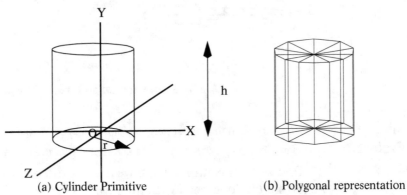

(a) Cylinder Primitive   (b) Polygonal representation

Fig. 2.18: The Cylinder model and its polygonal representaion.

A cylinder can be thought of as a disk whose perimeter has been stretched along the vertical axis by a distance h. This defines the vertical surface of the cylinder. The cylinder is capped off by a disk at its top. It is not necessary that either the top or bottom or both ends of the cylinder be capped off. In our software we will always define our cylinders to be capped on both ends. The polygonal mesh defining the cylinder can be constructed in a manner similar to that used for the disk.

Mini Project

- Go back to the model program and create a cylinder. You will get a form that will allow you to control the height and radius of this object. Create a cylinder of radius 20 and height 40 units.

- Change the complexity of the cylinder (the complexity option is under the **Draw** menu.) Notice how the polygons representing the cylinder increase in number for increasing complexity.

Cone

As with the cylinder, the cone can be defined by its radius (r) and height (h), and the location of its origin (0) (which is again at the world origin) as shown in Fig. 2.19. Some examples of a cone are an ice cream cone (the sugar ones), and a birthday party cap.

The cone is closed at the bottom by a disk representing its base. The polygons defining its vertical surface are lined along the perimeter of the base and meet at an elevation of h, vertically above the origin, as shown in Fig. 2.19.

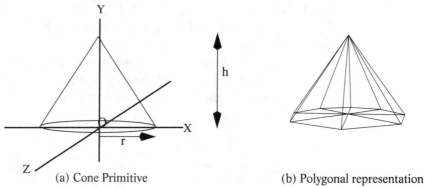

(a) Cone Primitive   (b) Polygonal representation

Fig. 2.19: The Cone model and its polygonal representaion.

Mini Project

- Create a cone in the model program with radius of 25 and height 40 units.
- Change the complexity of the cone and notice how the model changes.

One word of caution: while we discuss these models. In this book we are mainly concerned with the surfaces of the models, not with the interior. Our models are hollow so to speak, and the polygons define the crust of the model. There are other times when computer graphics people are concerned with the volume of the object. Such volumetric visualization (as it is called) is beyond the scope of this book.

The Generic Model

So far we have been discussing the primitive objects that can be defined mathematically. What about objects that are more generic in their representation? It turns out that if we define a sufficient number of polygons (increase complexity of a model), we can represent almost any shape. The more component polygons we define, the more accurate our model will be. Look at Fig. 2.20. We have shown some really complex models

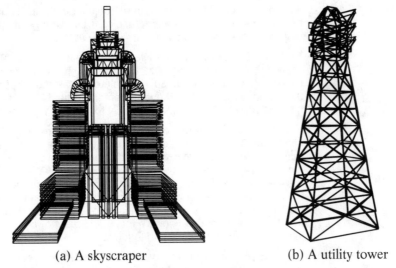

(a) A skyscraper   (b) A utility tower

Fig. 2.20: Generic models constructed from polygons.

that use more than a hundred polygons to define their surfaces! Can you imagine actually calculating each vertex of each polygon manually? It would probably take you the rest of the year to finish even a simple model of an ant! Luckily for us, there exist ways to generate these polygons, without figuring out the location of each polygon vertex. We shall look into them in the next section. Our software does not provide you with the tools to build these objects. However, we do provide support to read in various kinds of model files generated by professional modeling software.

Most professional software store model information in a file using specific formats; *dxf* is a popular format from AutoDesk. You can get dxf files by downloading them from various Internet sites. You may have to be careful; some dxf files do not specify the vertices of their polygons in an order to ensure back face culling.

We also have our own model file format called *mdl*, which we can read and write out using the software provided with the book. We have provided some information in Appendix B on the format of the mdl file. Enterprising readers can define their own model files, either by typing it in by hand or by running a simple program to generate mdl files. Let us try and load up a few of these models.

Mini Project

- In the model program, choose the **Open Model** option, under the File menu. A dialog box will pop up prompting you to set the directory and the file you wish to open. Change the folder to **C:\3dcg\Models**. The file selector lets you select two kinds of files, dxf files and our own mdl files. Change the selector to read Model files(*.mdl) files.

- Click on teapot. This will load up the ubiquitous Utah Teapot model, which has become a symbol of computer graphics. Observe how the polygons are defined to give you the impression of a smooth curvature.

- To read a dxf file, under the File menu select the **Open Model** file option and change the file selector to DXF (*.dxf) files. There are some dxf files provided for you also. First, go ahead and delete all the models from the world. To do this, choose the **Delete All** option from the **Objects** menu. Alternatively you can hit the **Ctrl** and **X** keys simultaneously to delete all the models.

- Load the skyscrpr.dxf model. This model has been built with more than a thousand polygons. If we used even more polygons, we would get a closer representation of the model. However you may already have noticed a significant lag in the time taken to redraw the model on the screen! You will also notice that sliding the complexity slider will have no effect on these models. This is because these models have been predefined with a certain set of polygons. No mathematical description is given for these models, so the computer does not know how to calculate and generate a more dense set of polygons. This is one problem inherent with dxf and mdl file formats.

- We have provided you with a bunch of models you can play with. Try loading them up and seeing how the surface is being defined through the use of polygonal meshes. You can even take a look at the file text to see how these polygonal meshes are being constructed.

2.5 Drawing Objects on the Screen

In this section we will understand how we display 3D objects on the computer screen. The screen is flat and it has no depth associated with it. How can we make it display objects that exist in a 3D world?

The goal of computer graphics is to give the viewer the impression that he/she is looking at the photograph of some three-dimensional scene. The visual output that is seen on the screen is the image that is seen by our CG camera (hence we called this window, the Camera View window in the model program). This camera has no physical existence, it merely exists as a few defined parameters which specify where it is positioned, and where it is looking. Look at Fig. 2.21. It shows a 3D scene and a camera placed in this scene. When the camera is *clicked*, an image of the scene is produced. The CG camera works in a similar manner and dynamically displays images of the computer-generated 3D world that it is viewing. Let us explore this further.

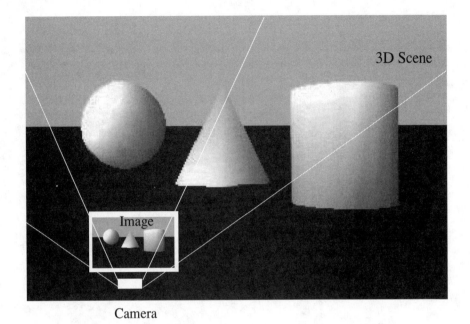

Fig. 2.21 The Camera, the Image, and the 3D Scene.

Consider what happens when you look at a photograph. A photograph is a two-dimensional replica of a three-dimensional world. However it still has some 3D qualities, which makes it seem 'real.' By looking at the two dimensional image we get certain visual clues of the 3D world due to some inherent qualities of nature. The qualities are

1. Objects further away from the camera look smaller than those objects that are closer (also called perspective view).

2. If we spin objects around, they have surfaces defining all three dimensions of their shape, i.e. they are 3D objects having depth.

3. As we move the object toward and away from the camera, the size of the object seems to grow and shrink.

Since we are trying to simulate the way a camera records an image of the world, let us try and understand the working of a simple pinhole camera.

Understanding The PinHole Camera

Fig. 2.22 shows a pinhole camera. The pinhole camera is a simple box, with a flat piece of photographic film (F) at the back and a small pin sized hole, H, which allows light in when opened.

Consider what happens when we take a photograph of object X. We open hole H for a fraction of a second. In this time, a ray of light hits the object at point P, passes through the pinhole of the camera, and

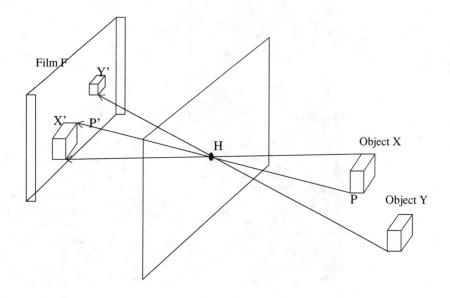

Fig. 2.22: The Pinhole camera concept.

hits the negative at point P', causing the film to be exposed at this point. The point P' forms the image of point P on the negative. All points on the object are mapped onto the negative in this fashion, and we get the image X' of object X, as shown in the figure. Another object Y, of the same size but further away from the camera, gets mapped on as Y', and is smaller in size. From Fig. 2.22 it may also be noticed that the pinhole camera actually creates an inverted image of the object in front of the pinhole.

The classic computer graphics version of the pinhole camera places the plane of the photographic film in front of the pinhole and renames the pinhole as the eye, viewpoint or camera position, and the photographic film as the image plane, as shown in Fig. 2.23. The display is placed in front of the camera for convenience in programming and modeling in computer graphics. This also ensures that the image captured is not inverted. Note how, in the CG version of the pinhole camera, each element of the pinhole system has an equivalent. The condition that only light rays passing through the pinhole and falling on the negative get captured on the film is equivalent to the condition that in the CG world, only light rays passing through the image plane and reaching the eye/ camera will be displayed. The image seen on the screen or image plane depends on where the eye is located and also what the eye is looking at.

As discussed above, images of objects in the 3D world are projected onto the two-dimensional image plane. This is the photograph that is

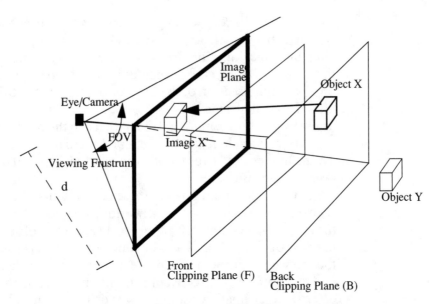

Fig. 2.23 Displaying a 3D object on a screen.

then mapped onto the display window on our computer monitor. If we were to draw lines from the eye to the corners of the image plane, the lines would form the edges of walls that include the eye and the image plane (see Fig. 2.23). This defines a view volume. Only the objects directly in front of the camera and within the walls of the view volume will get displayed on the image plane. These conditions mean that the 3D world that appears on the screen lies within an infinite pyramid with the base cut off (or a baseless pyramid). This 3D view volume visible to the camera or eye is called the viewing frustum. The angle formed at the tip of this frustrum is called the field of view of the camera, or the zoom. The field of view works similar to the zoom feature in a normal camera. Changing the field of view (or FOV) causes objects being displayed to zoom in or out on the display. Increasing the field of view (or the angle of the frustrum) will lead to a wider angle camera, causing the image to look smaller and hence more distant. Decreasing this angle will result in a telephoto like image, with the object zoomed up close in the display.

In most cases the view volume needs to be finite; that is, we need to limit the number of 3D objects seen in the viewing frustum. This is achieved by defining a front clipping plane (F) and a back clipping plane (B), as shown in Fig. 2.23. These planes are parallel to the image plane. Any objects that lie before the front clipping plane will not be seen, i.e. they will be clipped. Anything behind the back clipping plane will also be clipped. If an object intersects any of the clipping planes, then only the portion of the model will be drawn that is inside the clipped view volume. In Fig. 2.23 the image of object X is seen as X' on the display screen. Object Y will not be displayed, as it is being clipped by the back clipping plane. Limiting the view volume in this manner helps to eliminate unwanted objects from being displayed and helps the viewer focus on a particular section of the 3D world. Note that the final image displayed is a wire frame image. At this point, we assume all objects have a wire frame representation. When we reach the rendering chapter, we shall see how each object can be represented as a solid shaded model. The technique used to capture the image and display it on screen, however, remains the same.

Let us understand how all these components of the camera tie in with the image that we see of the world. We explore these concepts with the aid of a schematic view of the world as if were observing the world perched upon the Y axis and looking down. This is sometimes called the top view. Another concept we make use of is called the *bounding box* of an object. This concept is useful in depicting the extents of an object in a schematic view of the world. The extents of an object refer to the maximum and minimum of the X, Y, and Z dimensions of the object.

■ Mini Project

- If you have exited from the model program, please restart it. If not, go ahead and delete all the models from the world.

- Create a sphere model with a radius of about 25 units. Under the Draw menu select **BackFace Culling** to turn on the backface culling option.

- Under the **Show** menu, toggle the **Schematic View** option on. You will now see a new window, which we shall refer to as the schematic window, as shown in Fig. 2.24. This window is a schematic that shows you a bird's eye view of our 3D world, as if the bird were viewing it from way on top (along the Y axis) and looking down. This is also called the top viewThe camera view window will move to the left to make room for this window..

- In the schematic view you will also notice the bounding box of each object. If you toggle the **Names** option under the Show menu item, you can see the names of the objects in the schematic view. You can turn off the names by toggling the Names option to off.

- In the schematic view the camera itself is represented as a red cone (which is actually the viewing frustrum). The tip of this frutrum is the location of the camera. Anything outside of this frustrum will not be seen. The front and back clipping planes are also seen as red lines crossing over the frustrum. The image plane can be seen as another red line in between the two clipping planes

- When you start model, the camera is defaulted to be located slightly off the positive Z axis and looking down at a slight angle to it. This camera position lets you fully appreciate the perspective of the 3D world.

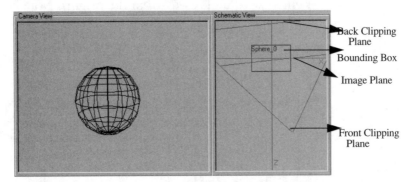

Fig. 2.24: The Top View of the 3D World.

In the software we have provided with this book, we have adopted the convention that the eye position (or camera location) is always at a fixed distance (d=200) from the image plane. We have provided the flexibility to move the front and back clipping planes. We shall see how to manipulate these components of the camera in the next section.

Now that we know how to create and display 3D objects, let us explore how to move these objects, and our camera in our 3D world.

2.6 Transformations

Once we have built our models, we may want to move them around in our 3D world to create a composition or a scene. We may even want to orient an object differently, or scale its size up or down. We may also want to angle our camera at different positions to yield different results. The functions used for modifying the size, location, and orientation of either the model or the camera are called geometric transforms. The most widely used transforms are translation (moving the model/camera location), rotation (changing the model/camera orientation), and sizing (changing the dimensions of the object not used for camera). Let us first explore transformations for objects.

Object Transformations

Translation

Translation is when we move an object along any of the three axes. Essentially this moves the local origin of the object. We can move the object left to right (translate it along the X axis), move it up and down (translation in Y), or move it forward and back (translation in Z). The abbreviations for these translations are Tx, Ty, and Tz. In Fig. 2.25 the box object is translated along the X axis from the origin to its new position (which is 10 units away from the origin). Note how this changes the location of the local origin of the object.

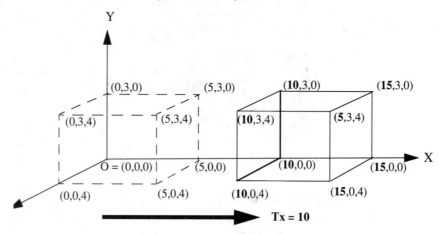

Fig. 2.25: The box model translated by 10 units.

Mini Project

- If you have model program running, toggle the **Schematic View** option to set it off from under the Show menu. This will eliminate the schematic view window from the last section. If not, restart the program.

- Click the Show menu and toggle the Axes option on, if it isn't on already. This will display the world coordinate axes in our Camera window, making it easier for you to appreciate which axes the different transforms are taking place about.

- Now create a box with dimensions roughly about length = 45, height = 25, and width = 40. Note that when you create the box, the local origin of the box is at the world origin.

- Under the Objects menu, choose the **Transform** option. This will bring up a form that will allow you to modify the position, orientation, and size of the object.

- Ignore the rotation and scale sliders for the moment. Use the TranslateObject sliders. These are sliders for translating the object along the X axis (Tx), Y axis (Ty), and Z axis (Tz). Each of these slides have positive and negative values to translate the object on both sides of the origin. Slide Tx around and watch your box move its location left and right of the origin. Similarly, Ty and Tz will move the box along Y and Z axes. A combination of moves along any of these three axes can navigate the model to any point in our world. Let us say we want to position the box at point P = (2,20,-5). You will need to move the box 2 units in X, i.e, set Tx = 2, 20 units in Y; Ty = 20, and -5 units in Z; Tz = -5.

- If you want to reset the object back to the origin, then click the Reset button and all transforms will be lost. Click the Reset button at this point.

Rotation

When an object spins, yaws or pitches, we say it has undergone a rotation. A spin rotates our model about the Y axis, and is called Ry. Pitching the model rotates it about the X axis, and is called Rx. Yawing the object is rotating it about the Z axis, and is hence called Rz.

Mini Project

- Now in the transformation form, use the Rx, Ry, and Rz sliders to orient the box in different directions. Note how the combination of rotations about X, Y and Z are enough to specify any orientation in space.

Rotations occur about a set of axes. By default, our objects rotate about the world axes, which intersect at the world origin. This means that rotation about the X, Y, or Z axes rotates the model with the world origin as a pivot point. Since we initially define the local origin of most models to be located at the world origin, this means that the object would simply rotate about itself. However, if we were to first translate the object and then rotate it, the local origin would no longer be coincident with the world origin. In this case, the object would revolve around the world origin when rotation is applied but not about itself. Let us try this.

Mini Project

- Click Reset on the **Object Transforms** form. Translate the box along the X axis by 15 units by setting Tx = 15. Now try rotating the box by changing Ry and Rz; set Ry = 50, Rz = 50. Notice now that the box is not rotating about its own local origin. Instead it is revolving about the world origin, as expected! Click Reset again.

- Now first rotate the object, set Ry = 50, Rz = 50. Translate it by setting Tx = 15. Note how you get different results. Transformations are not commutative. This means that the order in which you perform the transformations operation is important to uniquely identify the final orientation of the model. It is important to understand which point is being used to rotate the object about. Play around with these two transforms till you feel comfortable with how these transforms are occurring.

What if we want our object to rotate about some other point in our world, not the world origin? Say we want the object to revolve about a point O', as shown in Fig. 2.26. O represents the world origin in this Figure.

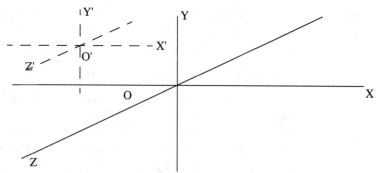

Fig. 2.26: The world axes, and the new set of axes defined at O'.

We can imagine a new set of axes, X',Y',Z' all orthogonal to each other, and parallel to the world axes. These axes intersect at O', the point in question. If we define rotations to occur about the new set of axes, X',Y' and Z', then the object will revolve about O' and not the world origin O. Let us see how to do this.

Mini Project

- Translate the object such that its local origin is now located at O'. Say we define O' = (-20,-30,0). Translate the box by Tx = -20, Ty = -30 so that the box's local origin is at this point.

- Now click the **Default Position** button on the **Object Transforms** form. This button instructs the software to use a new set of axes, X',Y',Z' which intersect at O' when applying rotations.

- Try rotating the object. It will rotate about the newly defined set of axes. If you move the model away from O' and then rotate it, it will continue to rotate about this pivot point.

- Another interesting feature that this button also supplies is that this new location is now defined to be the default position for the object. If you hit the Reset button for the box, it will default to this newly defined position, and not the world origin! This is a handy tool when you want to compose a scene with many objects located at different default locations.

- Reposition the object back to Tx=0, Ty = 0, and Tz = 0, and hit the **Default Position** button to re-locate this object about the world origin.

Scaling

The process of changing the dimensions of a model along any of the three axes (X, Y, Z) is called scaling. Scaling can also occur simultaneously along any two or all three axes. Sx, Sy, and Sz refer to scaling along the X, Y, and Z axes, respectively. Scaling essentially changes the size of the object by multiplying component polygon coordinates by the scale factor. In Fig. 2.27 we scale a polygon first in X, by setting Sx=2, and then in Y by setting Sy = 1.5. Notice how the scale factors affect the coordinates of the polygon.

Mini Project

- In your **Object Transforms** form click the Reset button. Now try scaling the box about the three axes, using Sx, Sy, and Sz. Set Sx = 2, Sy = 1.5, and Sz = 0.8.

- Like rotation, scaling occurs about a set of axes. If we first translate our object and then scale it, we would get a different result. Click on Reset. Set Tx = 15, and then set Sx =2, Sy =1.5, and Sz = 0.8.

Note that scaling, like rotation, depends on the set of axes about which scaling is occurring. By default it occurs about the world origin, and this is reflected in Fig. 2.27. If we change the default origin of the object to a new point O', (by translating it and hitting its Default Position button) then scaling will occur about the new set of axes X'Y'Z', which intersect at this new point O'.

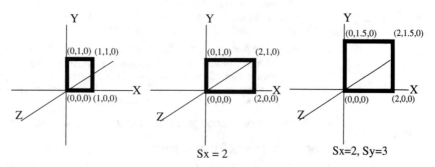

Fig. 2.27: Scaling a Polygon.

Mini Project

• Click Reset. Set Tx = 15 and hit the **Default Position** button. Then set Sx =2, Sy = 1.5, and Sz = 0.8. Notice that the box scales us about the point (15,0,0), the point where its local origin is currently located.

In general, when specifying the transformation applied to an object you need to specify the following:

1. The type of transformation
2. The axis along or about which the transformation occurs.
3. For scaling and rotation, the set of axes about which transformation occurs
4. The order in which a specified transformation will occur in a series of transformations

As you have seen, the primitive transformations can be combined or concatenated in different ways to get different end results. The proper sequence of transformations can only be achieved after the location of the pivoting point is identified. Later on we shall see how we use transformations to achieve animation.

For those mathematically curious, we shall provide a brief description of what is happening when we apply object transforms. Each object has associated with it what we call an object transformation matrix. This matrix is nothing but a set of numbers that identifies what transforms have been applied to the object. We saw in a previous section how to define the polygons of an object when it is located about the world origin. When we finally draw the object to the screen, we multiply the transformation matrix of the object to every vertex of every polygon of which the object is composed. This results in a new set of transformed polygons. When we display these polygons on the screen, the resultant image will show the object as having been transformed by the transformations applied to it.

Mini Project

• Create new primitive objects.

• Under the Draw menu, click the **World** option to toggle the option to see all the objects you have created so far. Note that one of the objects will be highlighted in maroon color. This is the currently picked object. The transforms affect only the currently highlighted or picked object.

- To pick an object, go to the Pick option under the Objects menu. A list of objects that you have created will be displayed, with each object identified by its type and a unique identification number. Select any of these objects to pick them. Any transforms you carry out will affect only the picked object. Do the transformations behave the way you want them to?

- Pick each object and move it around to compose a scene for yourself.

Camera Transformations

Just as it is possible to transform objects in space, similarly we can also transform our camera. This means that not only can objects be moving in our 3D world, but our camera could be moving with respect to the objects also. Transforming the camera changes the look of the entire world as expected. In the section on displaying 3D objects, we discussed how in our software the distance of the image plane or the screen from the camera is fixed. Another way to look at it is that the eye and the image plane move (translate and rotate) together as a unit.

In the Mini Project that follows, we will make use of camera transforms and also see how clipping planes affect the image displayed (see earlier discussion of front and back clipping planes). The camera also has a field of view (or FOV) that defines its zoom. When you started up the model program, the default position of the camera was slightly rotated and translated in X and Y. This position gave a perspective view of the models being constructed. For reference, the default position of the camera has the following values: Rx = 10, Ry =-9; Tx =38, Ty =40.

Mini Project

- We assume you have the model program running. If you are restarting the model program at this point, create a few models for yourself so we can experiment with them in this session. Turn on the World option under the Draw menu, and toggle the **BackFace Culling** option on so that back face culling is enabled.

- Under the Show menu, toggle the **Schematic View** option on. You will now see the schematic top view of the world. The schematic will help you appreciate how the different components of the camera affect the image being displayed on the screen. Recall how the different components of the camera are represented in this view from the last section.

- Now click **Transforms** under the **Camera** menu item. This will bring up the **Camera Transform** form.

- This form lets you manipulate the translations and rotations of our camera. When you started model, the camera was defaulted to a location at Rx = 10, Ry =-9; Tx =38, Ty =40, FOV = 53. This camera position lets you fully appreciate the perspective of the 3D world.

- If you click Reset, the camera sets itself at (0,0,0), FOV = 53 and looks straight along the Z axis. Click OK.

- Now try transforming the camera. Translate in X, Y, and Z. The schematic will show you what the camera is doing and you will observe the results of the transforms in the Camera View window. When you transform the camera by changing Tz, you will notice that you zoom closer or further away from the model. Since the image plane is at a fixed distance from the camera, when the camera is moved closer to the models the image formed on the image plane appears bigger. When you translate the camera in X, you are effectively moving the image of the models to the left and right with respect to the image plane. You may note that translating the camera in Y will have no effect on the schematic view. This is because we are viewing the world along the Y axis and only changes in the X-Z plane can be observed from this vantage point.

• Next try rotating the camera about the different axes. The camera always rotates about itself. Rotating the camera about the X axis pitches the camera up (or down), causing the image of the model to move down (or up). At a certain angle you will notice that the images can no longer be seen, as the objects are no longer in the viewing frustrum range of the camera. If you continue to pitch the camera the objects will once again enter the viewing frustum as the camera goes through a complete rotation.

• Rotate the camera about the Y and Z axes. Notice how the images appear to move with the camera movements.

• You can also try zooming the camera by changing the FOV scrollbar in the Camera transformation form. This will zoom the images in and out on the display screen. The viewing frustrum enlarges or shrinks its extents when its FOV is being changed. This causes objects to fall in and out of the camera's view.

• Let us also experiment with the front and back clipping planes. Click the **Clipping Planes** option under the Camera menu. This brings up sliders to control the front and back clipping planes. Watch the schematic window as you change the clipping planes. When the clipping planes start intersecting the object bounding box, you will see your object getting clipped in the modeling window! Clipping planes are used to cull out unnecessary objects that are too small or too close to be seen, but waste display time as the computer still tries to draw them. Effective use of clipping planes can lead to faster display of models.

• Reset the camera transforms by clicking the Reset button. You can hit the OK button to remove the **Camera Transform** form.

Again for those curious, a peek behind the scenes. We define what is called a camera transformation matrix, which identifies what transforms have been applied to the camera. When drawing the objects, the component polygons of the objects are multiplied by both the object transformation matrix and the camera transformation matrix to produce the final image of transformed objects. Don't let all these details confuse you. We are just giving you a glimpse of what is happening behind the scenes. This will in no way affect your understanding of the rest of the book.

Camera transformations are used a lot in motion ride simulators. If you have been on the *Back to the Future* ride in Disneyland, you know what it feels like to move through a 3D world. If the ride has too many rotations you can end up feeling nauseated too! In the animation chapter we will see a simple example of how to build the graphics ala motion ride.

2.7 Hierarchy of Objects

Three-dimensional objects can be grouped together to create structures that define how these models are transformed and how they relate to one another. Grouping of these models creates structures called hierarchical structures, because within these structural groupings some objects are always more dominant than others. The components of the structure are commonly referred to as nodes.

The objects at the top of the hierarchy are called parents; the objects below the parents are called children and grandchildren. The top level of the hierarchy has no parent. Children inherit their parent's transformations. This means that if a parent object is translated, its child object also undergoes the same translation. Hierarchical structures can also be visualized as an inverted tree structure where the highest level of importance in the structure corresponds to the trunk of the tree. The branches that come out of the tree are the next level hierarchy; branches coming out of the main branch are the next hierarchy level, and so on, till we get to the leaves. These links ensure that the spatial position of the children update when the parents are transformed.

As the user, you need to define this hierarchy. The best choice of hierarchy is ideally one that takes into account the movement/animation of the scene. Let us try to understand the concept of hierarchy by building our own hierarchical model.

Consider building a snowman—we shall refer to him as Snowy, shown in Fig 2.28.

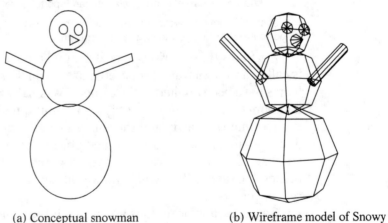

(a) Conceptual snowman   (b) Wireframe model of Snowy

Fig. 2.28: The conceptual snowman, Snowy.

Let us first consider the component objects that build up the snowman. The base, tummy, and head are of course all spheres. The hands are cylinders, the eyes are disks, and the carrot-shaped nose is a simple cone.

How would we define the hierarchy of such a model?

We definitely want the eyes and the nose to ride along with the face. This would mean we would define them to be children of the head object. Similarly, the hands can be grouped as children of the tummy. Now the base, tummy, and head need to be parented in a way that would enable us to move them as a single unit. In order to do this, we define what we call a *Null Node*. This object is used solely for the purpose of creating a hierarchy. It has no representation in our world. It is a virtual object that has a location, a local origin that by default is equal to the world origin, and can be transformed. The introduction of such an object would lead us to a hierarchical structure of the snowman as shown in Fig. 2.29.

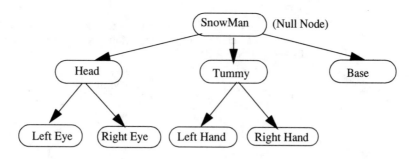

Fig. 2.29: The Snowman model hierarchy.

Mini Project

- If you have exited from the model program restart it. If you have model running from the last session, go ahead and delete all the models from the world. Hide the schematic view by turning off the **Schematic View** option under the Show menu.

- Create a sphere with a radius of 18 units.

- Let us rename this object as **Base**. To do this, select the **Name As** option under the Objects menu. A form pops up prompting you for the name of the object. Clear out the existing name and type in **Base**. Click OK to set the name.

- Toggle the World option under the Draw menu on, so we can see all the models we are creating. Also enable back face culling so we don't see the rear of the objects. Create two more sphere's of radius 12 and 8 and rename them **Tummy** and **Head** respectively. When renaming, be sure that the object you want to rename is the one that is currently picked and hence highlighted.

- We shall now create the null object. The null object is, by default, located at the world origin. Under the Objects menu, select the Create option and click on the **Null Node** option. This object has no representation and will not show up in the Camera View window. Rename this object as **SnowMan**.

- We want to group the three spheres to be children of the SnowMan object. Under the Objects menu, select the **Group** option. Fig 2.30 shows the form that will pop up.

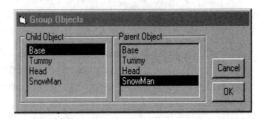

Fig. 2.30: The Group Objects Form.

- This form lets you pick two objects, and set up a grouping between them. One of the objects is the parent and the other the child. Pick Base as the child object, and pick SnowMan as the parent. Click OK to create the link between these two.

- Click the Group option again and create links to set up first the head and then the tummy as children of the SnowMan object. We have now created the hierarchy shown in Fig 2.31.

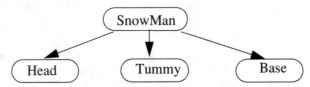

Fig. 2.31: The hierarchy created.

- We have created a hierarchy, but each individual child needs to be placed at its correct default location in this hierarchy. We want the local origin of the entire group to be at the tip of the base object. The null object, being the parent, defines the origin about which transforms for the entire group will take place.

- Pick the Base object and translate it up in Y by 18 units, so it is located at the origin of the null object. Hit **Default Position** and then click OK.

- Pick the Tummy and translate it up in Y by about 44 units, so it sits on the base (there will be a slight overlap between the base and the tummy). Hit **Default Position** and then click OK. This will set the Tummy object to have its default position at Y=44. Next pick the head, and translate that by 61 units, so it sits (slightly overlapping) on the tummy. Set this as its default position.

- Now pick the SnowMan object, and try moving it around. You should see the base, tummy and the head moving as a unit. These three objects are now grouped together! If you pick the tummy and transform it around, the other objects will not follow! This is because the links go only in one direction; only the children follow their parents, and not vice versa. Hit Reset to reset the tummy back to its proper position.

- Now create two eyes by creating two disks. Create a disk of radius R = 2. Note that when you create the disk object, you may not see it because of its size and orientation. But we will take care of it in a moment. Rename it as **LeftEye**. Link it as child of the Head object. This will cause it to be positioned at the center of the head. Move the LeftEye out to Tz = 8 and Tx = -3 so it is positioned at the surface of the head. Hit the **Default Position** button to set this point as its default location. Rotate it about X by -90 so the LeftEye object faces you.

- Create a disk of radius R = 2. Rename it as **RightEye**. Link it as child of the Head object. Move the RightEye out to Tz = 8 and Tx = 3 so it is positioned at the surface of the head. Hit the **Default Position** button to set this point as its default location. Rotate it about X by -90 so the RightEye object faces you.

- Similarly, create two cylinders with a radius of 2 and height of 20 units. Rename them as **LeftHand** and **RightHand**. Link them as children of the tummy, and set them to have a default position of Tx = +/- 9. Rotate them about the Z axis so Rz = +/- 50 so that they are positioned at the sides of the tummy, as shown in Fig. 2.28.

- Create the nose by creating a cone of radius 2 and height 7. Make it a child object of the head, and position/orient it as shown in Fig. 2.28 by setting Ty = -4, Tz = 8. Hit Default Position. Set Rx = -115 so the nose points down a little. Name it as **Nose**.

- Now if you pick the Snowman object and transform it about, all objects will follow its transforms. You will note that there is no red highlighted object, as the object picked (the null node snowman) has no representation. The children will transform about the pivot point of the parent. If you pick each object individually, they will transform about their own local pivot point. Keep this in mind while playing around with the snowman.

- Try squashing and stretching the snowman by changing Sx and Sy appropriately. Try spinning him around. Try waving his hands up and down. Be sure to pick the object you want to transform.

- Save your work by choosing the 'Save Model' under the File menu, and saving the model with the name **snowman** in your **C:\3dcg\Models** folder. This will save the model as **snowman.mdl** file. We shall reload it later on to do all kinds of fun things with the model.

- We have provided a backup **snowman.mdl** file for you in case you lose yours. You will find this file in the **C:\3dcg\Examples** directory.

(Note that saving models only saves the translation information and not the orientation or scaling information. If you read in the snowman model later on, you will have to reset these values.)

Now that you are familiar with most of the concepts of modeling an object, we provide an exercise for you to do on your own to test your understanding. We encourage you to pick up objects you encounter in your daily life and experiment with building these objects using the software provided.

Exercise

Try and create a christmas tree whose components objects are as shown in Fig 2.31.

As you can tell, the christmas tree is made up of a cone, with a cylinder as the trunk. The spheres form the ornaments of the tree. The star is itself made up of component polygons. You will have to create these polygons, and then group them so that they form a single unit. You will

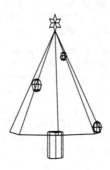

(a) The Christmas Tree

(b) The component objects of the Christmas Tree

Fig 2.31 The Christmas Tree and its component objects

want to group the tree appropriately to make it behave as a single object. Save your work when you are done, so you can re-load your model whenever you want to.

2.8 Advanced Modeling

Although the surface polygonal approach provides a convenient method for constructing objects, it can be rather tedious when building complex models. Various shortcuts have been developed to aid in the building of certain generic forms. To determine the orientation of the polygon, these programs often let the user point and click to define front and back facing polygons. We describe some of the techniques here. These are advanced techniques and our software provides no support for such modeling methods.

Extruding

Extruding constructs a boundary representation of a model by sweeping a user-defined cross sectional shape along an axis to create an extrusion with a given length. In Fig. 2.33, we show a cross section and the 3D surface created by displacing it along the Z axis.

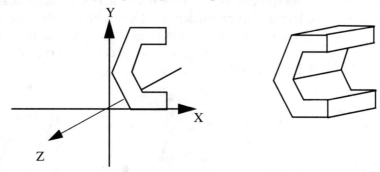

Fig. 2.33: A 2D shape extruded along the Z axis.

The extrusion can also take place along some predefined curved path to define more complex extruded shapes.

Surfaces of Revolution

Surfaces of revolution help in constructing objects like glasses, bottles, and in general any object having symmetry about a central axes. A master 2D contour input by the user is taken and rotated about one of the three axes to sweep out a surface of revolution. The final surface is still constructed from planar polygons. Fig. 2.34 shows how a wine glass can be modeled by revolving a 2D contour.

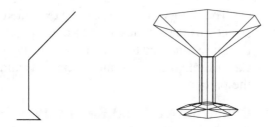

Fig. 2.34: A wine glass created by revolving a 2D contour.

Cubic Curves and Surfaces

So far we have been dealing with straight lines to depict curved surfaces. More complex mathematics is used to represent smooth curves and surfaces. These curves are controlled by a designated set of points also called as *control points,* which indicate the general shape of the curve.

These control points are used as the basis for polynomial functions that define the mathematical equation defining the curve. Typically, the polynomial equation is cubic in nature (of order 3), and hence these curves are called cubic curves. Cubic curves are more easy to control than higher-order polynomials, while at the same time provide a smooth continuous path. The control points act as a kind of magnet that pulls the curve in certain directions. Depending on the type of cubic equation, the curve may or may not pass through its control points.

Some of the common types of cubic curves, with their control points labelled are shown below.

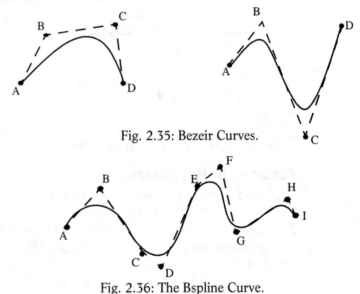

Fig. 2.35: Bezeir Curves.

Fig. 2.36: The Bspline Curve.

The theory of curves can be extended into two dimensions to construct surface patches. Once a cubic patch is defined, these patches can then be connected to form a mesh to generate a smooth curved surface. Care is taken to maintain a seamless and smooth boundary between the patches.

Constructive Solid Geometry

Constructive Solid Geometry (CSG) attempts to build objects by adding and subtracting objects from basic defined shapes to generate more complex objects. For example, in the Fig 2.37, we subtract model B from model A to get the CSG-generated model C.

Digitizers and Scanners

The simplest and perhaps most accurate way to generate a model is to scan it in. A laser scanner works by shining a beam of light on the model. Sensors detect the variation in the geometry of the model as the laser

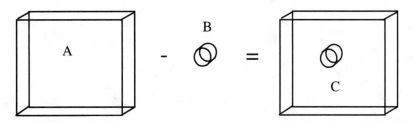

Fig. 2.37: CSG Model.

pans across the surface. The scanner digitizes the important features of the model and records them as data points. These points can then be used by the software to generate component polygons. Sometimes, if the model is made of plaster or clay, you can even actually draw a network of polygons and then use a hand held probe to identify each vertex.

While many more ways to generate models exist, it may be obvious to you that all techniques cannot encompass every situation. At different times, we may need a different set of tools to create our model. The objective of all these different techniques is to get the characteristic features of the model input into the computer.

2.9 Summary

In this chapter we discussed the process of modeling as it is used in computer graphics. The XYZ coordinate system is commonly used to define the three-dimensional space in which objects are modeled. Using vertices and lines in this space, we were able to define basic shapes such as a polygon, a box, a cylinder, a cone, and a sphere. A simple polygon-mesh model to represent more complex shapes was also discussed. Using our basic shapes and polygon approach, we were able to model Snowy the snowman, and also build our own Christmas tree. We also covered some common techniques for building more complex models. This should provide you with a strong background in the theory and practice of computer modeling

The pinhole camera concept helped us understand how 3D models are displayed on a computer screen. We also learned how to translate, rotate, and scale a modeled object, as well as transform the camera. We encourage you to experiment with the software as much as you can till you feel comfortable with creating models and transforming them to orient them in a desired position. The concept of transformations is the key to unlocking the mysteries of animation, and we shall use it extensively in the next chapter.

CHAPTER 3
Animation

3.1 Introduction

To animate means to bring to life. In the last chapter we saw how we can use principles of computer graphics to model objects. Our next task is to put life into these modeled objects.

When you see a movie on television or in the theater, you see characters in motion. This motion is usually smooth and continuous. The actual movie film or video tape containing the movie, however, consists of a sequence of images or frames. These frames when flashed in rapid succession on the screen cause the illusion of motion we see in the movie. Fig. 3.1 shows such a series of frames from an imaginary filmstrip where a smiley face changes its smile to a frown.

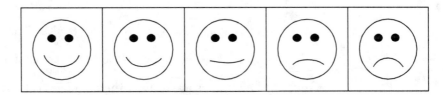

Fig. 3.1: A Sequence of frames in a filmstrip.

The illusion of continuous motion seen in movies exists because of a peculiarity in the human vision system. When you see a picture, it tends to remain in your mind's eye for a brief period of time. This is called persistence of vision. If the pictures are shown quickly enough, then a new picture arrives before an old one fades and you perceive a smooth transition between the frames. If the difference between successive frames is too large or the frames are not shown quickly enough, then the illusion shatters and the movement appears jerky. This kind of effect is seen in some of the older movies and documentaries. Typically a film needs 24 *frames per second* (also called *fps*) to look smooth and continuous. In the case of video, 30 frames per second are needed

As in movies, an illusion of motion in the computer graphics world is achieved by playing back a sequence of images/frames at a certain speed. The frames are created by defining modeled objects in different positions and maybe even different orientations in three-dimensional space at different times. By playing back these frames at 30 frames per second, we can form an animated sequence. A lot of techniques of traditional two dimensional animation to create and modify frames are used in 3D animation. We will begin by learning about traditional animation as an important background reference.

3.2 Traditional Animation

In traditional 2D animation, most of the work is done by hand. The story to be animated is written out on a storyboard. Initially, the *key frames* are identified. *Key frames* are frames selected by master animators for their importance in the film, either because of the composition of the frames, an expression depicted by characters, or an extreme of movement in the sequence. Consider the smiley face sequence we saw in Fig. 3.1. For a typical animation, the key frames of this sequence would be frames with the smile and final frown, as shown again in Fig. 3.2.

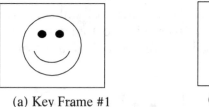

(a) Key Frame #1   (b) Key Frame #2

Fig. 3.2: Key frames in the smiley animation.

Once the key frames are drawn, the next step is to draw the intermediate frames. The number of intermediate or in-between frames that are drawn depends on the frames needed to give the illusion of smooth motion in a specified time. This step of drawing in-betweens is called *tweening* or in-betweening.

If we were going to create a video of the smiley animation, we would need to create 30 frames per second to get the illusion of smooth motion. So doing some simple math for a short 5-minute animation we would need [frames per sec] x [sixty] x [length of time in minutes] = 30 x 60 x 5 = 9000 frames! That means 9000 frames to be hand drawn by animators. Now you know why Disney hires so many animators to work on a single movie. Traditional animation is definitely a time-consuming and labor-intensive process. How can the animation tasks be made less laborious and also more flexible? The answer is found in the use of computers for animation.

Computers are used by traditional animators to help generate in-between frames. The animators sketch the key frames, which are then scanned into the computer as line strokes. The computer takes the initial and final frames and generates the necessary strokes to produce the in-between frames. The use of computers in animation is not limited to this minor role. Since the advent of computer graphics, computers are being used in a big way to generate the entire animation sequence in a simulated three dimensional world. Using sophisticated algorithms and equations, the in-betweening can be made to look very natural and smooth. In the next few sections we will look at some fundamentals of computer animation. Using the software, in this chapter we will also be able to create some of our own animations.

3.3 Concepts in 3D Computer Animation

A wide variety of 3D computer animation techniques are currently used by various production houses. Depending on the desired effects, different techniques and tricks are used. Even though all these techniques are different , they are often combined to achieve a more appealing animation.

The most popular technique used in computer animation is the key frame interpolation technique, or in-betweening adapted from conventional animation. Two or more key frames are identified by the user. The challenge is to smoothly reorient and move objects or models generating a sequence of in-between frames to create an illusion of motion. Computing the in-between frames by averaging the information in the key frames is called interpolation. The interpolation creates a sequence of still frames. Depending on the type of equation used to average the information, we get different kinds of interpolations.

The interpolation techniques can be used to calculate the position of objects in space, as well as their size, orientation, and other attributes. Given a set of key frames, in-between frames are created by calculating the position and shape of objects at that frame using the interpolation equation. Each in-between frame differs slightly from its adjacent frames and this process is continued till we reach the next key frame. In the discussion that follows we use the time and frame number of an animation interchangeably. Both these parameters are related to each other as shown below:

30 frames = 1 second of video or,

24 frames = 1 second of film

Each key frame has a set of parameters like position (x, y, z coordinates) and orientation associated with the models in the frame. Interpolations are usually carried out by varying one or more of these parameters between key frames.

Each key frame has a unique position on the animation timeline. Note that we borrow the term *frame* from traditional animation. It is a term that has become very common among 3D animators also, and is essentially the image of the world at a given time along the animation timeline.

Let us learn about interpolation techniques and how the transformation parameters are interpolated between key frames.

Linear Interpolation

The simplest type of interpolation is linear interpolation. In linear interpolation the values of a parameter are estimated between two known values, assuming that all values lie on a straight line (hence the term linear). In simpler terms, linear interpolation averages the parameters in the key frames and provides as many equally spaced in-between frames as needed.

Consider an object with its local origin at an initial position P_{in} = (X_{in}, Y_{in}, Z_{in}) at key frame KF_{in}, which then attains the final position of P_{fin} = $(X_{fin}, Y_{fin}, Z_{fin})$ at keyframe KF_{fin}. Say we wanted only one in-between frame. This frame would be midway between KF_{in} and KF_{fin}, as shown in Fig. 3.3a, and would have the object midway between positions P_{in} and P_{fin}. The coordinates of the object at this in-between frame F_i, given by (X_i, Y_i, Z_i) would be:

$$X_i = X_{in} + (X_{fin} - X_{in})/2 = (X_{in} + X_{fin})/2$$

$$Y_i = Y_{in} + (Y_{fin} - Y_{in})/2 = (Y_{in} + Y_{fin})/2$$

$$Z_i = Z_{in} + (Z_{fin} - Z_{in})/2 = (Z_{in} + Z_{fin})/2 \qquad \text{Eq. 3.1}$$

Now suppose we needed two in-between frames, F_i and F_j. The object would then be defined between P_{in} and P_{fin} such that at F_i, its position (X_i, Y_i, Z_i) is one-third away from P_{in} and at F_j its coordinates

(Xj, Yj, Zj) are one-third away from P_{fin} and two-thirds away from Pin. In total we have 4 frames: KF_{in}, Fi, Fj, KF_{fin}, shown in Fig.3.3b. The coordinates at the two in between frames would be calculated as follows:

$$X_i=X_{in}n+(X_{fin}-X_{in})/3 \qquad X_j=X_{in}+2*(X_{fin}-X_{in})/3$$

$$Y_i=Y_{in}+(Y_{fin}-Y_{in})/3 \qquad Y_j=Y_{in}+2*(Y_{fin}-Y_{in})/3$$

$$Z_i=Z_{in}+(Z_{fin}-Z_{in})/3 \qquad Z_j=Z_{in}+2*(Z_{fin}-Z_{in}) \qquad\qquad \text{Eq.3.2}$$

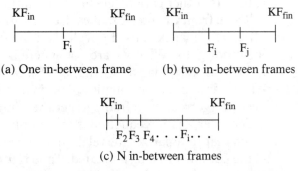

(a) One in-between frame   (b) two in-between frames

(c) N in-between frames

Fig. 3.3: The in-between frame positions in the animation timeline.

If we were to extend this, in general if we wanted a total sequence of N frames (including the initial and final keyframes, hence N-2 in-between frames), then the ith frame, F_i would have coordinates (X_i, Y_i, Z_i) given by:

$$X_i=X_{in}+(i-1)*(X_{fin}-X_{in})/(N-1)$$

$$Y_i=Y_{in}+(i-1)*(Y_{fin}-Y_{in})/(N-1)$$

$$Z_i=Z_{in}+(i-1)*(Z_{fin}-Z_{in})/(N-1)$$

where i goes from 1 to N.   Eq. 3.3

If we set i = 1 in the above equation we get the initial point P_{in}, and for i = N, we get the final position, P_{fin}. As we go through frames i=1 through i = N, the position of the object is linearly interpolated between P_{in} and P_{fin}. In the above case we assumed that the first frame of the sequence was frame 1. However, this may not always be so. We can get a more general equation for the interpolation between two key frames, resulting in a sequence of N total frames and starting from initial frame number I, as:

$$X_i = X_{in} + (i-I)^* (X_{fin} - X_{in})/(N-1)$$

$$Y_i = Y_{in} + (i-I)^* (Y_{fin} - Y_{in})/(N-1)$$

$$Z_i = Z_{in} + (i-I)^* (Z_{fin} - Z_{in})/(N-1)$$

where i goes from I to I+N-1   Eq. 3.4

In the above examples we used the position of the object (or the X,Y, and Z coordinates of its local origin) as parameters for interpolation. It is also possible to apply the same type of linear interpolation equation for other parameters like the rotation and scale of the object.

Recall from Chapter 2 that each object has its own transformation matrix. This matrix is used to calculate the final position of its component polygons. What we are doing here is calculating the transformation matrix associated with the object at each in-between frame using the linear interpolation equation. This matrix is then multiplied to the object polygons at each frame to get their intermediate positions.

The interpolation equation is commonly expressed in the form of a graph that summarizes the relation between time (or frame number) and the parameter being animated. Time is usually represented on the horizontal axis, and the animating parameter, on the vertical axis.

Note that we use time and frame count interchangeably, as both are directly related to one another, viz. 30 frames (or 24 for film) is equivalent to 1 second.

Bouncing Ball Example

In this section we will look at an example that is commonly used to explain animation principles — the bouncing ball. We will simplify the animation tasks so that it is easier to grasp concepts. We will initially begin by looking at movement of the ball along only one axis and we will work with wire frame objects.

As shown in Fig.3.4a, a ball is initially at position A, with its origin located at (0,A,0). It falls to the ground at position B, which sets the ball's origin to (0,B,0). The ball finally bounces back to position C, which is lower than position A, and the ball's origin is at (0,C,0). The ball moves vertically (Ty along the Y axis) and its movement is graphically represented in the accompanying graph in Fig. 3.4b. The vertical axis of the graph is the y-displacement of the ball and the horizontal axis is the time axis. This graph also has a guide to show where the ball actually is in the display window in the three key frames.

The movement of the ball can be split into two stages:

stage 1: ball goes from A to B, and

stage 2: ball goes from B to C.

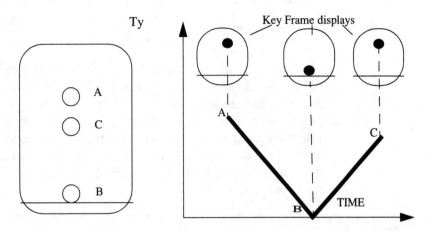

(a) The key frames of the bouncing ball   (b) Graph of Y position of ball vs. Time

Fig. 3.4: Linear Interpolation for a bouncing ball

The positions A, B, and C are called key frames, as they are key in defining the flow of our animation. We need to figure out how to find the positions of the ball in the intermediate frames. Since the ball does not move in the X and Z axis, the motion along these axes will always remain zero and we will not need to calculate it. We only need to calculate its position along the Y axis.

Let us say we want this animation to be 20 frames long, with the ball starting at frame 1 at position A, reaching position B at frame 10, and reaching position C at frame 20.

For linear interpolation between position A and position B, we go through 10 frames, from 1 to 10. The initial frame is 1 and the last frame is frame 10. We want to find out the interpolated value for the Y position of the ball at any intermediate frame (between 1 and 10). Using Equation 3.4 from the earlier discussion, total frames from A to B, N = 10, initial frame I = 1, the interpolated value of Y at any frame f between 1 and 10 can be found as:

Intermediate Y Position (bet A & B) at frame $i = Y_i = A + (i - 1)*(B-A)/(N-1)$

$= A + (i-1)*(B-A)/9$

as i goes from 1 to 10.   Eq. 3.5

Note how in Eq. 3.5, at i=1, the intermediate position =A, at i= 10, position =B, as expected.

Similarly, we can calculate intermediate positions between B and C as (note initial frame I=10) :

Intermediate Y position (between B and C) at frame i=Y_i=B+(i-10)*(C-B)/ 9

as i goes from 10 to 20.   Eq. 3.6

Using Equations 3.5 and 3.6, we are now able to find the position of the bouncing ball at any intermediate position. This is a simple example but the same principles can be applied to more generic linear interpolations. So much for the math behind interpolation. Let us turn our attention to some exercises. This will help you grasp the concepts better. We will make use of the user-friendly interface provided with the animation package to learn and experiment with interpolation techniques using graphs.

Mini Project

- Start the animation program by clicking on the animation.exe icon. Refer to Appendix C on how to start the program.

- You will see an interface similar to the model program, with an additional menu item, **Animation Options**. Also note that some of the forms and options may differ from the ones in the modeling program to some extent. The software has been modified slightly to enable you to define up to 60 frames of an animated sequence.

- We will begin by creating a sphere model of our bouncing ball. Click the Create->Sphere option under the Objects menu to create a sphere with a radius of about 15 units.

- Next select the **Object Transform** option from the Objects menu. The Object Transforms form that pops up has some extra sliders and buttons to let you edit in different positions for the ball at different frames. The **Current Frame** slider indicates which frame you are currently viewing. The other sliders work the same way as we saw in the modeling chapter.

- The Current Frame should be frame 1. Drag Ty to 50. You will see the ball being drawn high up in the Camera View window. Click the **Save Key** button. This button will save the frame as a key frame, with the transformations you have set; in this case Ty = 50. This is position A, as shown in Fig. 3.3.

- Position B is set in frame 10. Note that point B is on a ground plane on which the ball will bounce. We will assume that the ground plane is at Y = -50 (this choice is made to get a full view of the ball bounce in the Camera View window). Slide the Current Frame slider to 10. Translate the ball down in Y to 50, so the ball is drawn low down in the window. Click the **Save Key** button.

- Next the ball bounces back up to position C in frame 20. Edit in values for position C by setting the current frame to be 20, and setting Ty to a value less than A, say around 30. Remember to save the frame as key frame, or our changes will have no effect on the animation.

- Now we have set all our key frames. Click OK in the Object Transforms form.

- We have three key frames at frame numbers 1, 10 and 20. The computer's task is to calculate what the position of the ball will be in the in-between frames, and to draw the ball at the appropriate location.

- At this point let's take a look at the graph we have created for this animation. Under the Show menu, pick the **Interpolation Graph** option. This will bring up a window to the right side of our Model window. This is the graph window. You should be looking at the graph of Tx against frame number. Since we have not changed Tx, this is simply a straight line along the X axis. Change the **Animated parameter** option to Ty by clicking on the option arrow and selecting Ty. You should now see a graph, as shown in Fig. 3.5. The key frames will be clearly marked with the red circles.

- Note how for frames beyond position C, the graph draws the same Ty position as at point C (the straight line after point C). This is because we have not identified any key frames beyond point C and by default, all frames beyond point C will have the same parameters as the key frame at C.

- Next select the **Animation Options** menu and choose the **Set Frame Number** option. This slider will let you change the current frame of the scene in the display window (similar to the frame slider we saw in the Objects Transforms form). If you move this slider to change the frame number, you will see a blue circle on the graph showing you the location of the current frame number and value of the animated parameter on the graph. In the model window, you will see the ball following the path defined by this graph.

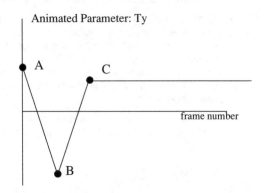

Fig. 3.5: Linear Interpolation Graph of key frames.

- Note how by changing the frame number, the ball is drawn at its interpolated position using parameters from the key frames. Our next task is to create the animation sequence and watch it play smoothly. This is done by saving each frame as an image and then playing these images in a sequence. To do this, click the **Animation Options** menu and choose the **Set Frame Range** option. This will bring up a form which will let you set the range of frames to be used for our animation sequence. The initial frame is 1. Set the final frame to be 20. Click OK. Note that the maximum number of frames we allow is 60.

- Click on the **Animation Options** menu and select the **Make Animation** option. This option will build the animation sequence. This option can also be selected by simply hitting the **F5** key on the keyboard. You may have to wait for some time while the animation is made and saved to disk. A form will pop up to inform you when the animation is ready.

- When making and playing back the animation, the graph window will disappear as it only hinders the speed of display.

- Once the animation is built, a control panel will appear on top of the display window, as shown in Fig. 3.6. The control panel allows you to play the animation, step forward a single frame, and loop through the animation. To make the ani-

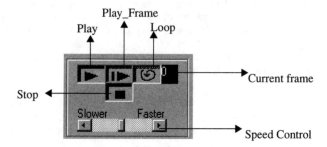

Fig. 3.6: The Control Panel.

mation play continuously, select the Loop button in the playback frame. Note that the only way to stop the Loop playback is by clicking on the Stop button. Also the loop playback stops only after it has played a full sequence of images after the Stop button is pressed. The current frame indicator only shows up when you are stepping forward one frame at a time

• Click on the Play button from the Control Panel to play the animation. The animation will display all objects in black (that is, there will be no highlighted object). Depending on the speed of your computer, you should be able to see a smooth animation sequence being played at a certain speed. In order to increase or decrease the speed of the playback, select the Playback Speed scroll button. Moving the speed control allows you to control the speed of the animation played.

•   In the next section of this project we will make the ball bounce further by adding in new key frames, D, E, F, and G as shown in Fig. 3.7

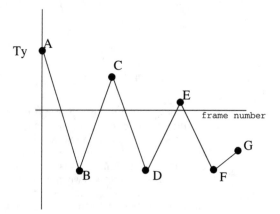

Fig 3.7: Graph of Ty vs Time in the bouncing ball animation.

•   Make the animation 60 frames long by setting frame D at frame 30, and a Ty of -50, frame E at frame 40 and Ty = 5, frame F at frame 50 with Ty = -50, and finally frame G at frame 60 with Ty = -35. To do this you will have to again make use of the Object Transforms form under the Objects menu. Remember to hit the **Save Key** button when you are done editing a key frame. If you edit in an incorrect key frame, you can remove it, by going to that frame and hitting the **Remove Key** button.

• Next set the range of animation to be from frame 1 to frame 60. Click the **Ani-**

mation Options menu and select the **Frame Range** option. Set the last frame to be 60.

- Choose the **Make Animation** option and build the animation. Play the animation and notice how the ball bounces up and down a few times on the screen, gradually bouncing lower and stopping at point G.

- Next let us make the ball move horizontally across while it is bouncing. To do this, edit the key frames to have a varying Tx value in addition to Ty, which we just added. Set the following values

    pos A: Tx=-80 frame=1

    pos B: Tx=-30 frame=10

    pos C: Tx=15 frame=20

    pos D: Tx=40 frame=30

    pos E: Tx=60 frame=40

    pos F: Tx=85 frame=50

    pos G: Tx=100 frame=60

- Remember to save each key frame after setting each Tx value at the appropriate frame. Do not change Ty, as we have already edited in the values for it.

- Clicking the **Remove Key** button will remove the current frame from being a key frame for the picked object. You can use this button if you make a mistake when you define a frame to be a key in your animation sequence.

- If you want to see the graph associated with Tx, you will have to bring the interpolation graph up again and change the animated parameter option from Ty to Tx. You can change the current frame value up and down to observe the ball's motion and its corresponding graph.

- Make this animation and watch your ball bounce across the screen. Note how in this animation, both the X as well as Y values of the bouncing ball are being changed in each frame.

- Do no exit the program as yet. We shall use this animation to illustrate another kind of interpolation technique in the next section.

In the above animation exercise, you may notice that the motion of the ball at key frames is a bit abrupt, especially when the ball reaches extremes, i.e., its highest point and lowest point. At these points the ball changes direction almost instantaneously. If you have ever seen a real ball bounce, you know that it slows down at the peak before changing directions to fall down. Do not end the animation program at this point. We shall look into how we can solve this problem!

One key feature in linear interpolation is that if two adjacent frames have the same value, then the in betweens will also maintain the same value. Linear interpolation is based on constant speeds between key frames, but produces abrupt changes in speed between two adjacent key frames when one constant speed ends and another one starts. We see this in our exercise when the ball reaches position B and starts falling to C abruptly. This kind of interpolation cannot handle subtle changes in speed because the frames are created at equal intervals along the time line. We can handle more sophisticated in betweening by using non linear interpolation techniques.

Non-linear Interpolation

As we noticed in the earlier section, the use of linear interpolation works in simple cases. However, for most real situations to be animated, we need to make use of more complex non-linear interpolation principles to simulate effects in nature.

Non linear equations are of many types. The most popular types are the cubic family of non linear equations/curves such as *B-spline*, *Bezeir* and *Hermite* curves. The non-linear nature of the curve takes into account the variations of speed over time. We looked into some of these curves in the modeling chapter. The math equations explaining these curves are beyond the scope of this book but interested readers should read the references listed in the bibliography.

A graph representing non linear interpolation is also called a function curve or a parameter curve. The key frames are used to define the control points of the interpolating curve. Several definition of curves exist depending on the algorithm used to draw a smooth curve to fit the key frames. In this book we make use Hermite curves to interpolate the key frames. To draw Hermite curves the key frame locations define control points through which a smooth curve is drawn. In order to define a Hermite curve we need at least four key frames.

In the next mini project we will look at the example of our bouncing ball but this time using hermite graphs.

Mini Project

- Assuming the animation program is still running, click on the **Animation Options** menu. Select the **Interpolation Technique** option and change it from linear to non-linear. Make sure your graph window is visible.

- Notice how the linear graph of our bouncing ball is now replaced by a curve, as shown in Fig. 3.8. This is the defining interpolation graph. As you can see, the graph curves at the top and bottom. This means the ball will actually slow down before changing its direction.

- Using the **Frame Number** slider, see how the ball slows down before changing directions at the key frames.

- Make the animation and play it. Watch the ball smoothly move from A to G.

- Note how the ball slows down before it reaches a key frame. This kind of slowing down before changing direction is desirable only at the peaks. When the ball hits the ground plane (at B, D, and F), this slowing effect is actually unnatural. We want the ball to pick up speed just before hitting the ground. Using non-linear interpolation is not a sure solution to simulate all real life situations. A solution to the problem is to let the user edit the interpolation curves to the desired shape. More sophisticated software for animation will let the animator tweak some of the control points to change the shape of the interpolation curves.

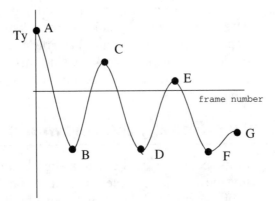

Fig. 3.8: Non Linear interpolation for the bouncing ball.

- There are other cases when cubic interpolation does not work too well. Consider a situation where in key frame C we want the ball to remain on our imaginary ground at Ty = -50. Bring up the Object Transforms form and slide the Current Frame slider to frame 20 (key frame C). Set the value of Ty = -50 at this frame and save the frame as a key. Now look at the graph of Ty versus frames. Notice that the graph between B and C, and also between C and D, wiggles below these points, as shown in Fig. 3.9. If B, C, and D were on the ground

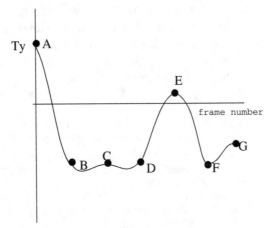

Fig 3.9 The curve wiggles below the control points!

plane, then the ball would go right through the ground. We would need to add extra control points to eliminate this undesirable side effect. Unlike linear interpolation non-linear interpolation does not retain the values of adjacent key frames when they are equal.

- You can now exit the animation program by holding down the **Ctrl** and **D** keys simultaneously.

In the above discussion of interpolation principles, we saw how depending on the desired animation effect, different schemes can be applied. It is hard to achieve accurate realism in all animation sequences. In most cases a combination of different schemes is needed to get good results. Extensive, ongoing research is being conducted to formulate new and better ways to make more realistic interpolations possible. In any production, the animator has a big role in understanding how interpolation or tweening works. Using this knowledge, a suitable interpolation technique needs to be adapted for a particular animation sequence.

3.4 Animating Snowy, the Snowman

In this section we will create a slightly more complicated animation using the techniques we learned earlier. For the animation we will make use of the model of the snowman we built in the modeling chapter. If you have not built the snowman, you can pick up the snowman.mdl we have provided for you in the *c:\3dcg\Examples* directory. We will make use of the concepts of hierarchical modeling and transformations we studied in the last chapter.

In the animation sequence we will put life into Snowy, our snowman, who will bounce up and down on a plateau. The plateau model will be a simple box object primitive. The animation will be about 40 frames long and Snowy's movements will be similar to that of a bouncing ball, wherein Snowy will bounce up and down and also move across the display screen. In addition to this we will also tilt Snowy back and forth to convey the direction of his motion.

On with the actual animation ....

Mini Project

- Start up the animation program. From the File Menu select the **Open Model** option and load the **snowman.mdl** file saved from the modeling chapter.

- We want a plateau for the snowman to bounce on. Load in the **plateau.mdl** file in the **c:\3dcg\Models** directory. Let us transform it. Select the Transform option under the Objects menu for this object. Set Tx = -11, Ty = -88, & Tz = 46. Also scale it out in x by setting Sx to 1.2. Leave the other transformations as is. Remember to save your key frame when you are done or else the transformations will not be saved. Hit OK

- Under the Draw menu, select the World option. This will draw all the component objects of the snowman as well as the plateau. Also turn the back face culling option on to see the models more clearly.

- We learned in the modeling chapter that the mdl file does not save the orientations of components. So the snowman model will have its eyes, nose, and hands oriented incorrectly. We need to reset these orientations.

- From the Objects menu, select the Pick option and pick the nose object. The nose component of the snowman model will be highlighted. Next select the Object Transforms option and orient the nose by setting Rx to about -106. Click the **Save Key** button to save this orientation.

Note: When you are setting the orientations for the snowman model, make sure the frame number is set to be 1.

- Next pick the left eye, LeftEye. you may not see the LeftEye component high-lighted as it is oriented parallel to the camera view. Set Rx = -90 at frame 1 for the LeftEye. When you do this the LeftEye object should become visible. Do the same for the RightEye and set Rx = -90. Save the key frame. By saving these orientations as key frame 1, we are making sure that the selected objects retain these transforms for the initial frame and will not change till altered in the next key frame.

- Pick the left hand and yaw it outwards, by picking the LeftHand object and set-ting its Rz to about -50. Save this frame. Pick the right hand, RightHand, and set its Rz to about 50. Save this keyframe.

- We also angle our camera to get a better view of the animation. Click the Trans-forms option under the Camera menu and set the camera Tx = 7, and Ry = -4. Save this key frame by clicking the Save Key button.

- Pick the SnowMan object. Remember this is a null object and will not have a display associated with it. However, the SnowMan object is a virtual object with its center at the lower base of Snowy. So any transformations applied to i will cause all its children objects to be transformed about this point.

- Bring up the transformations form by choosing the Transforms option under the Objects menu. Set the Current Frame to be 1. Set Ry to -50. This will orient the snowman so we see his profile. We shall also translate the Snowman to be drawn high up and left on the screen, and tilted backward to convey that he is moving forward while he is coming down. To do this, set Tx = -82, Ty = 24 and Rz = -20. Save this key frame.

Note: Remember to save your frame after you are done transforming the object for a frame. Also if we do not change a particular transfor-mation for a key frame, then assume that it retains its value from the last key frame.

- At frame 10, we want the snowman to hit the plateau. Slide the Current Frame slider to frame 10, and set Tx = -24, Ty = -40. As mentioned above, leave the other transforms as they are and save the frame.

- At frame 13, we shall pitch Snowy to make him upright. Slide the Current Frame slider to 13, and set Rz = 0. Save the frame. At frame 16 we shall pitch Snowy forward getting him ready to bounce back up. At frame 16 set Rz =20. Save the key frame.

- At frame 26, Snowy shall reach the highest point in his trajectory. At this frame set Tx = 16 and Ty = 24. The other transforms stay the same as before. Save the key frame.

- At frame 30, Snowy will be pitched backward to start his downward journey. Set Rz = -20. Note that all transforms being applied to the component objects of Snowy are being applied around the null object called Snowman. So when we pitch Snowy back, his base remains at the same point. In this case, we want Snowy to actually move forward, to convey the sense that he is rotating about his tummy while in mid air. To do this, set Tx = 44. Save the key frame.

- At frame 40, Snowy has just reached the level of the plateau surface. He is oblivious to the fact that the plateau does not continue till this point. Set Tx = 84, and Ty = -40. In this section we shall create an animation of only 40 frames. So we are done setting all our key frames.

- To set the frame range for animation, click the **Set Frame Range** option under the Animation Options menu. Set the Final Frame to be 40. Make your animation.

- Play back the animation and watch Snowy bounce up and down and also move across the plateau.

- You have just completed your first successful animation task. Save your animation by clicking the **Save Anim** option under the File menu. Save your animation as **snowman** (snowman with plateau.). Also save the model as the same name by choosing the **Save Model** option under the File menu.

- When you save the model and anim file, remember to give them the same names. The next time you load an animation file by choosing the **Open Anim** under the File menu, it will search for the model file with the same name and load it up also. Although you are saving both animations and model files with the same name, we append a ".mdl" or ".anm" to identify it is a model file or an animation file. If you happen to glance at your directory using Windows Explorer, you will see this extension added to the end of the file. When you are ready to start with the next section, you can reload your animation by choosing the Open Anim option and choosing the snowman animation file. This will also load up the correct model file for you. We have provided you with a backup animation file, called **snowmanwp.anm** in the **c:\3dcg\Examples** directory in case you ever need to make use of it.

Now that you have animated your first sequence its time to do some post-analysis. The Snowy animation worked well but you will agree that the bouncing Snowy did look a bit stiff and unnatural. Also the animation lacked any sense of drama and appeal. So what do you think we did wrong? The answer to some of our problems can be found in principles of traditional animation. Some fundamental principles of classic 2D animation that have been used in traditional animation for over 50 years help to make animations look real and appealing. Understanding these principles is essential to producing good traditional as well as computer generated animation. In the next section we will acquaint ourselves with some of these principles and also apply them to our Snowy animation.

3.5 Principles of Traditional Animation (In the Realm of Computer Animation)

Three-dimensional computer animation has made great advances in the last few years. There are several simple, user-friendly, and flexible animation packages available that can help even a novice develop quality animation. However, it is quite common to see that some of the animations may show up as drab, computer-generated images. Here is where principles of traditional 2D animation become important. The concepts used in hand-drawn 2D animation can make animations more realistic and complement the benefits provided by computers in 3D animation. In this section we will look at some of the critical principles that will add zest and depth to our computer animations.

Squash and Stretch

A very important concept in animation is called squash and stretch. When an object is moved, the movement emphasizes any rigidity in the object. In real life, only the most rigid objects retain their original shape during motion. Most objects show considerable movement in their shapes during an action. One trick used in animation to depict the changes in shape that occur during motion is called squash and stretch. The *squashed* position depicts a model form either flattened out by an external pressure, or constricted by its own power. The *stretched* position shows the same form in a very extended or elongated condition.

The most important rule to squash and stretch an object in animation is that the object should always retain its volume. If an object squashes down, then its sides should stretch out to maintain the volume. The standard example of this principle is the bouncing ball. Fig. 3.10 shows some of the frames in a bouncing ball example. The ball is squashed once it has made impact with the ground. This gives the appearance of bouncing. The ball is elongated before and after it hits the ground to enhance the feeling of speed. The squash-and-stretch effects help to create illusions of speed and rigidity, and is a good technique to simulate realism. Human facial animation also uses the concepts of squash-and-stretch extensively to show the flexibility of the skin and muscle and also to show the relationship between parts of the face. We will use the principles of squash and stretch to add some life to our Snowy animation.

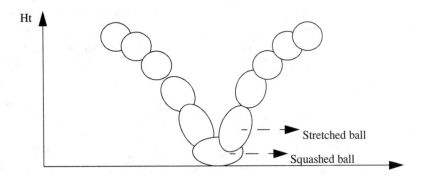

Fig. 3.10: Squash and Stretch applied to a bouncing ball.

■ **Mini Project**

- In this exercise we will apply the squash-and-stretch concept to Snowy. If you have the animation program running from the last section, you can continue right away. If not, please go ahead and reload the snowman animation you saved from the last session. You can do this by clicking the **Open Anim** option under the File menu and picking the **snowman** file.

- Pick the Snowman object and select its Objects Transforms option. We shall be using Sx, Sy, and Sz (scaling transforms for the Snowman object) to squash and stretch Snowy. At frame 1, we wish Snowy to retain his original shape. So we shall not edit this frame.

- At frame 10, Snowy is just about to hit the ground. Let us elongate him in this frame. To elongate him we need to stretch him out along the Y axis. Let us set Sy to 1.2. In computer animation, when we stretch an object along one axis we squash it in another dimension to maintain the volume of the object. In this example it is not possible to mantain the exact volume of Snowy. However, to create this visual effect, we shall flatten out his sides. Set Sx to 0.8 to do this. Save the frame once you have set these values.

- At frame 13, Snowy has made impact with the ground of the plateau. We shall squash him at this frame by setting Sx = 1.2, Sy = 0.8. Save your frame.

- At frame 16, Snowy is about to take off for another bounce. Set Sy to 1.2 and Sx to 0.8. Remember to save the frame once your transformation is edited in.

- At frame 40, Snowy is just about to land on thin air. We shall make Snowy stretch out as if he were landing on ground. Set Sx = 0.8 and Sy = 1.2 This addition adds a comic interest in the animation spot.

- We don't want to change Snowy for the rest of the animation, so we shall leave the rest of the frames as they are for now. Make your animation and play it. How do you feel about the animation? Does it add more life into Snowy's bouncing? Does Snowy have a more realistic bounce ?

- Do not exit from the snowman animation for the remainder of this section. If you do wish to exit the program, please save your animation so you will not lose your work.

Staging

Another important concept in animation is staging. Staging is the presentation of an idea so that it is completely and unmistakably clear. A personality is staged so that it is recognizable; an expression or movement is brought into focus so that it can be seen. It is important that only one idea be seen by the audience at a time. If a lot of action is happening on the screen at once, the eye does not know where to look and the main idea will be upstaged. The camera placement is very important to make sure that the viewer's eye is led exactly to where it needs to be at the right moment.

Let us apply staging to our Snowy animation. We shall make Snowy complete his second bounce only to find himself off the plateau and standing on thin air! To stage this part of the animation, we want to pitch the camera up and focus on Snowy. Such a swing in the camera is called a cut. A cut splits a scene into distinctive shots. The transition from shot to shot can be a direct cut, which we shall use in this example.

It could also be a smooth transition from the last frame in the first shot, to the first frame in the second shot. We shall look into such transitions in a later chapter. Usually the last few frames from the first shot are repeated in the second shot to maintain a continuity between the shots. Let us implement these concepts to refine our Snowy animation.

Mini Project

- In order to get a good camera position for our effects we will alter the camera transforms for the last few frames (from frame 41 onwards) in the Snowy animation. Before we do this we want to ensure that the camera remains fixed at its original position from frame 1 to frame 40. We have already set the camera at frame number 1. If we set frame 41 to have the new orientation, the camera will drift as we go from frame 1 to 41, as its transforms will also be interpolated. We will talk more about camera transformations in a later section in this chapter. For right now we need to ensure that the camera does not interpolate its position between frame 1 and frame 41. To do this we shall set frame 40 to have the same camera transforms as frame 1. In the Camera menu, select Transforms and slide the frame number to frame 40 and save the frame. This will make the camera transforms for frame 40 the same as those for frame 1. At frame 41 we want the camera to cut to the next shot. Slide the frame slider to 41 and set Tx = 50, Ty = -39, Tz = -12 and Rx =-14. Save this frame and click OK on the form.

- Pick the Snowman object. Bring up the objects transforms form. Slide the frame number Slider to frame 41. We shall set frame 41 to have the same position as frame 36, to overlap the actions of Snowy from shot1 to shot2. Move the frame slider to 36 to get the transformation values. Slide the slider back to frame 41 and set these values as Tx = 68, Ty = -14, Ry = -50, and Rz = -20, Sx = 0.9 and Sy = 1.1. (We have rounded off the scale factors for Snowy's stretch.) Notice the angle we are watching Snowy at now. It is more dramatic as the camera is viewing Snowy right at the center, leading our eyes to watch him and his expressions at a close range.

- At frame 43, set Snowy to have Tx = 78, Ty = -20, Sx = 0.8, and Sy = 1.2. Snowy has just landed. Save the key frame.

- At frame 46, we shall bring Snowy back to his original shape by setting Sx = 1 and Sy =1. We also straighten him out by setting Rz = 0. Save the key frame.

- Using the **Frame Number** option in the **Animation Options** menu slide the frame number from 1 to 50. Do you see how the shift in the camera viewpoint leads our eye to watch Snowy more closely as he lands on thin air? You can go ahead and build the animation and play it or wait until the next section to add more features. The tricks we used in this section illustrated some key concepts used in staging an action in animations.

Anticipation

Anticipation involves preparing the objects in a frame for the event that will occur in later frames in the animation sequence. If there is no anticipation of a scene, the scene may look rather abrupt, stiff and unnatural. The anticipation principle makes the viewer aware of what to expect in the coming scene and also makes the viewer curious of the coming action. Anticipation is particularly necessary if the event in a scene is going to occur very quickly and it is crucial that the viewer grasp what is happening in the scene.

Continuing our Snowy animation, we already have the stage set up for the audience. They have seen Snowy land on thin air. To create more anticipation, we shall make Snowy look down to give an impression that he is trying to figure out what he is landing on. This would draw a viewer's attention to the fact that good old Snowy is going to have a free fall from this frame onwards.

Mini Project

- From the Objects menu select Pick and pick Snowy's head. Till frame 45 we want the head to retain its original transforms. Bring up the Object Transformations form. Set the current frame to 45 and hit Save key. This will set frame 45 to the default orientation of the head. Since there are no key frames from 1 to 45 for the head, the head will retain this orientation till frame 45.

- At frame 52 ,we want the head to be pitched down. At frame 52, set Rx for the head to -50. Set Ry = 17 to swing his head towards the viewer. Save this key frame. Let's continue on and wait for the last few sections to view our complete animation in just a few moments.

Timing

Timing means the speed of an action. It gives meaning to the movement in an animation. Proper timing is crucial to make ideas readable. Timing can define the weight of an object. A heaver object is slower to pick up and lose speed than a lighter one. Timing can also convey the emotional state of a character. A fast move can convey the sense of shock, fear, apprehension, nervousness. A slow move can convey lethargy or excess weight.

In the Snowy animation, we let Snowy look down lazily, using 7 frames to look down (from frame 45 to frame 52). However, when he realizes his coming plight (free fall), we want him to look up in shock. We shall make him look up in only two frames to convey this sense of shock.

Mini Project

- With the Head still being the picked object, at frame 54, set Rx -2 and Ry =50. Save the key frame.

- To exaggerate the sense of shock we can even scale up Snowy's eyes at this point. Pick the LeftEye. Keep its original transforms till frame 50. To do this simply slide the Current frame slider to frame 50 and hit Save. At frame 54, scale up the eye by setting Sx = 1.6 and Sy = 1.6. Do the same for the right eye. Save the key frame. We're almost there now. Lets look at the last trick we can apply to our Snowy animation.

Secondary Action

A secondary action is an action that results directly from another action. Secondary actions are important in heightening interest and adding a realistic complexity to the animation.

For Snowy, a secondary action would be his hands moving up and down as he bounces up and down. This motion of his hands would help him maintain his balance and also add some realism to his bouncing. His hands would move down when he is landing and raise up as he bounces up. Can you add in these transforms to Snowy's hands?

■ **Mini Project**

- We will leave the secondary motion transforms to Snowy's hands as an exercise for the reader. Once you are done adding secondary motion, save your animation in a file called snowman.

- Make the animation and play it. Watch the animation and take special note of the tricks you have used in this section to make the animation more fun and realistic. Particularly notice the change in camera viewpoint, the roll of Snowy's head, his eyes widening, and his hand movements too.

- Try changing the interpolation technique for Snowy to be non-linear . What do you see? Is the motion better or worse than its linear counterpart?

- Please save your animation in a file called snowman. We shall be re-using this animation in later chapters to illustrate other aspects of computer graphics. We have provided you with a backup animation file called **snowmanwp.anm** in the **c:\3dcg\Examples** directory in case you lose yours.

In addition to the above mentioned techniques, there are several others that can be applied to computer animation. Exaggeration or deliberate amplification of an idea/action is often employed to focus the viewer's attention. Another common technique used for achieving a subtle touch of timing and movement is called slow in and out. In this the object slows down before reaching a key frame. We saw how this makes for a natural look of the bouncing ball when it reaches its highest point. The use of all these animation principles, along with good presentation of a theme, can help produce eye-catching and entertaining animation.

3.6 Viewpoint Animation

In the last few sections we learned some tricks to make our animations look more realistic and appealing. In all the previous examples the position of the camera or our viewpoint was fixed. We did change the camera position in the Snowy animation, but that was to present a better view of Snowy. We saw how the change in camera position led to a different shot in the animation sequence. Camera movement is not limited to changing across shots only. A technique often employed in animation is to move or transform the camera itself through a shot. This technique is called viewpoint or camera animation.

In viewpoint animation the camera is transformed through space using the camera transformation we described in the previous chapter. The transforms are interpolated exactly like the interpolation we use for object transforms. The camera is assigned parameters at key frames, and the intermediate frames get interpolated camera positions. For each in-between frame, the camera transformation matrix is computed using the same interpolation techniques we used for object transforms. This transformation matrix is then used (in conjunction with the object transformation matrix) to compute the final position of the object.

Why would one want to animate the camera? Well, camera movement adds valuable depth to the scene. The images we display are being viewed by you on a two dimensional screen. Although we apply tricks from photography to simulate the effect that you are viewing a three-dimensional scene, animations can tend to look flat. This is because we don't get a good three dimensional perspective on the spatial relationship of objects in the scene. Moving the camera through space causes objects to move in different ways depending on their distance from the camera, cluing us into how they are spatially arranged in our 3D world space.

Since the camera represents the observer's eye, a lot of planning is required to move the camera. The speed, field of vision, and orientation of the camera are critical because what the camera sees is what an observer would perceive. The faster the camera moves, the faster the observer feels he/she is moving. The sense of speed due to change in position and orientation obtained by camera motion can produce eye-catching and exciting effects. These techniques are commonly used in motion ride simulators where you feel like you are moving through a real-life scene. Other spectacular examples of viewpoint animation can be found in animations of the solar system, motion of atomic particles in a virtual model, and so on.

Let us explore the way a camera can be transformed in a shot. Let us start off with the snowman example itself. Recall how we defined an additional key frame for the camera transformation in order that it did not drift between the two shots we were defining. We shall now remove this key frame so we get a smooth camera motion between these frames.

■ **Mini Project**

- Assuming you still have the snowman animation running, bring up the Camera Transforms form. (If not, go ahead and re-start anim and load up the snowman animation you saved from last session).

- Go to frame 40. This is the frame we had added to the camera in order that it did not drift. Click the **Remove Key** button. This will remove this frame from being a keyframe. Now drag the current frame slider up and down. Do you see how the camera pans smoothly across with Snowy? Click OK

- There is still a bit of a jerk as Snowy animates across. This is because we had also defined an additional few frames of overlap for Snowy across the two shots. Let us remove these key frames also. Pick the Snowman object and bring up the Objects Transforms form. Go to frame 40 and hit the Remove Key button.

- Now go ahead and make your animation. Watch it play back. Do you see how the camera smoothly pans across with Snowy? Does it help the animation in any way? Already you get a sense of perspective between Snowy and the plateau he is jumping on. If there were more objects in the scene, the sense of perspective between the various objects would be even more keen.

The snowman animation was a simple example of a camera animation. Let us explore how we can achieve even more visually appealing results by looking into another example. For this viewpoint animation, we will make use of a model of a three-sided gallery as shown in Fig. 3.11. We shall move our camera through the gallery whose top is open to the sky. The top view of the gallery model is shown in Fig. 3.11a and the side view is shown in Fig. 3.11b. For the animation we will move, our camera through the center section of the gallery so that as we move we get to see both the inside walls as well as parts of the open sky. Some camera key frames of the camera's motion through the gallery are shown in Fig. 3.11a as cones (marked as points A, B, and so on).

On with the mini project ...

■ Mini Project

- Restart the animation program.

- Choose the **Open Mdl File** option under the File menu. Go to the **c:\3dcg\Models** directory and pick the **gallery** mdl file.

- When the gallery model first loads up, the camera position will already be inside the two walls of the gallery. The height of the walls is about 30 units.

- Bring up the Camera Transformation form by clicking the Camera menu and selecting the Transform option. At frame 1, set the camera Ty = 21. This will let your viewpoint be somewhere in the middle of the gallery walls. Set Tx = -65 and Tz = -60. This frame corresponds to key frame A, shown in Fig. 3.11. Save this frame.

- We shall keep Ty fixed at 21 for the rest of the animation, so there will be no need to change this parameter.

- At frame 6 the camera moves forward to key frame B. Change Tz to be -84 at this frame. Save the frame.

- Now we shall be rounding the curved area of the gallery. The camera will need to rotate yaw about itself to do this. In fact, it may be helpful to bring up the Schematic View window to see exactly how the camera is being transformed. Remember that the camera always rotates about itself. Rotating the camera about the Y axis will cause the camera to yaw around the gallery model.

- At frame 12, set Tx = -60, Tz=-119, and Ry = 21. We have just started rounding

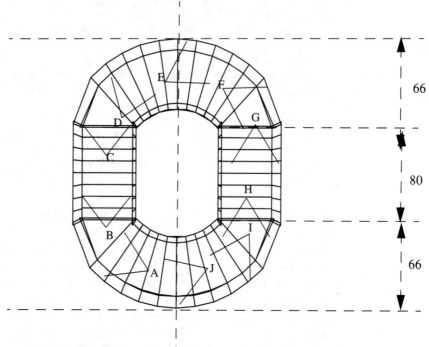

(a) Top View of the gallery and some camera key frames.

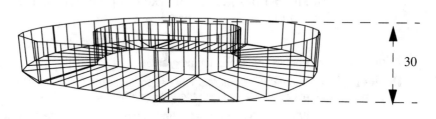

(b) A Perspective Side View of the gallery model

Fig. 3.11: The gallery model.

the corners. Looking at the Schematic window, will show how the camera is rounding the corners. The gallery itself will only be represented by its bounding box and not its actual shape. Save the keyframe.

- At frame 18, set Tx = -36, Tz = -152 and Ry = 60. Save the key frame.

- At frame 24, set Tx = 21, Tz = -170 and Ry = 127. Save the key frame.

- At frame 32, set Tx = 65, Tz = -139 and Ry = 180. We have now rounded the curved area and are back to traversing the straight section of the gallery. Save the key frame.

- At frame 37, change Tz to be -116. Save the key frame.

- The next few keyframes will be rounding up the second curved portion of the tunnel.

- At frame 43, set Tx - 60, Tz = -63, and Ry = 226. Save the key frame.

- At frame 49 set Tx = 27, Tz = -32, and Ry = 265. Save the key frame.

- At frame 54 set Tx = -28, Tz = -32, and Ry = 318. Save the key frame.

- Since we want to be able to loop this animation, we want the last frame of the animation to have the same location as the first frame. This will create a seamless loop when we play back the animation. So at frame 60, set Tx = -65, Tz = -60, and Ry = 360. Remember that a rotation by 360 degrees is a complete rotation and hence our camera will be back to where we started from!

- Make the animation and play it back. Does it look a little jerky? This is because the linear interpolation cannot result in a smooth rounding around the curved areas, as it results in a linear path from key frame to key frame. We shall have to add more key frames in this region for the animation to look more curved. Alternately, we can change the interpolating technique to a non linear interpolation to get a smooth well curved. Go ahead and make the animation with the interpolation technique set to non-linear. Does the animation look smoother especially when we are rounding the curved areas?

- Save this animation in a file called gallery. We have provided you with a backup animation file called **gallery.anm** in the **c:\3dcg\Examples** directory.

Camera animation gives the illusion of flying through space with you as the pilot of the flight! In fact, that is exactly what we are doing. Our camera is flying through our computer graphics scene, and the images being seen by the camera are being displayed on the screen. This gives a much different effect compared to the objects themselves flying through the scene. Depending on the animation being done, you can combine object animation and camera animation to achieve appealing effects. As an exercise, try to enhance the gallery animation by adding in new objects that come flying towards the camera and make the camera dodge these objects.

There are other tricks used for camera animation. One is changing the zoom of the camera. This is used often when one wants to open an animation to give the viewer a sense of the 3D world and then zoom in on the character in focus. Some of the more advanced animation techniques use entirely different approaches to animating a sequence of frames. We shall discuss some of these techniques in the following section.

Exercise

Try animating Peter, the hero on the cover of our book, on a skateboard. We have provided some models in the *C:\3dcg\Models* directory. Peter is in the peter.mdl file, hills.mdl contains a hilly terrain model, and board.mdl contains the board portion of the skateboard. You will need to add wheels to it .

(Hint: wheels could be made out of simple cylindrical objects).

Enjoy!

3.7 Advanced Animation Techniques

As we mentioned in earlier sections, computer animators use a variety of animation techniques. Some of these techniques are quite different from those used in the traditional key-frame approach. Depending on the type of effects needed for a particular animation, several techniques have evolved over the years. Generally, film makers combine two or more of these techniques to achieve a final animation sequence. We discuss some of the advanced animation techniques briefly in this section.

Dynamics

Dynamics is the branch of physics that describes how objects move. It uses the physical properties of objects (like mass, size) and the properties of the world that they live in (like air resistance, gravity, etc.) to calculate the animated path the object would traverse. The virtual world is hence imbued with simulated forces modeled after the laws of physics in the real world. The forces acting on the models cause action to occur. For example with this technique, a ball would be let go in a virtual world having a gravitational force, and this force would dictate the bouncing of the ball.

A dynamic system tries to make the motion as realistic as possible. These kinds of simulations are used widely in flight simulators and other animations where real-life effects in nature need to be closely replicated.

Procedural Motion

When you want a specified kind of behavior or action, you can write a program that makes the action for you depending on some input parameters. You use a set of rules to animate your scene. Hence, control of motion is achieved through the use of procedures that explicitly define the movement over time. The most popular of these methods is the particle system. Fuzzy objects like fire and clouds can be represented by particle sets. These particles are simple-shaped objects, like spheres or cubes. The set of particles evolve in space and die at different times. The procedures control the position, velocity, and also the color and size of the particle. A collection of hundreds of particles can lead to very stunning visual effects of realistic fire and explosions.

Motion Capture

Real-time motion capture is an advanced technique that allows animators to capture live motion with the aid of a machine. An actor is strapped with electronic nodes that transmit signals to the computer when the actor moves. This enables the computer to record the actor's motions. This captured motion data can then be applied to computer-generated characters used in animation.

Usually the animation achieved by such a technique needs to be enhanced using other animation techniques to clean out noise and irrelevant data due to hardware limitations.

Kinematics

Kinematics is the study of motion independent of the underlying forces that produced that motion. Kinematics is important to study articulated figure animation which has become very popular due to the increasing desire to use human beings as synthetic actors in 3D environments.

An articulated figure is a structure that consists of a series of rigid links connected at joints. In computer animation, these joints can be rotated to cause the attached links and joints to rotate about it. There is a hierarchy involved in the chain of links and joints. Rotation about any joint causes all the children to rotate but leaves the parents untouched. There may be some constraints in how much each joint will allow the link to rotate about each axis. This corresponds closely to the human anatomy, and hence this study is applied to most human animations.

Consider Fig. 3.12a where we show three links L1, L2 and L3 with joint J1, J2 and J3. J1 is the head of the hierarchy, and L3 is the at the bottom of the hierarchy. A rotation about J1 causes links L1, L2, and L3 and joints J2 and J3 to rotate, as shown in Fig. 3.12b. Similarly, a rotation about J2 will cause L2, L3, and J3 to rotate.

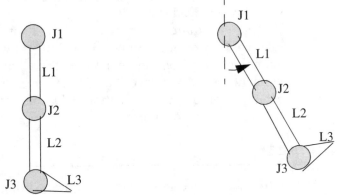

(a) The Links and Joints chain   (b) Rotation about Joint J1

Fig. 3.12: Rotation of a Linked Object.

We can compare this organization of joints and links to the human leg. J1 is the hip about which the thigh, L1, rotates. The knee corresponds to the joint J2, which rotates the calf, L2, the ankle joint J3, and the foot L3. You can see that any study of the motion of the model in Fig.

3.11 will provide a good tool for computing an animation of a human leg in motion. Most parts of the human anatomy can be represented as such skeletons whose motion can then be studied by the use of kinematics.

In forward kinematics, the motion of all joints is specified explicitly by the animator. The motion of the links is determined indirectly as the accumulation of all transforms that lead to the link as the tree of the structure is descended. In the case of our leg, the foot (L3) would move as a combined effect of the rotation about the hip (J1), the knee (J2), and the ankle(J3), as shown in Fig. 3.13.

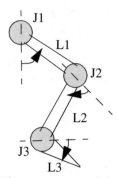

Fig. 3.13: The movement of the links in the hierarchy.

In inverse kinematics, the animator specifies the final position of the links at the end of the tree structure. Math equations are then used to determine the position and orientation of all the joints in the link hierarchy that lead to this link.

There are quite a number of other animation techniques that are used regularly by different people in different walks of life. Computer animation has become an integral part of all presentations and entertainment media. We have provided an introduction to concepts and techniques of computer animation in this chapter that can form the basis for more advanced studies.

3.8 Summary

In this chapter we looked at techniques to bring our 3D models to life by animating them. We looked at the classic example of a bouncing ball. Linear interpolation techniques provide a good starting point to animate sequences. But to add realism to the animation, we need to make use of nonlinear or cubic interpolation.

3D animation can be very drab if we do not apply principles of traditional animation to them. We looked at squash and stretch, timing, anticipation, and staging techniques to add life to our animations. The project to animate Snowy, our friendly snowman, gave us an opportunity to apply some of these tricks. In section 3.6, we learned about viewpoint animation and how it can help to give a new feeling of depth to our animations as we move along with the camera.

And just like in all the other chapters, we have just touched on the basics needed to unlocking the secrets of computer graphics and animation. Using the subroutines for linear and nonlinear interpolations, and several other animation routines, you can create your very own animations. In the next chapter we will focus on adding color, textures, and lights to our scenes (called rendering).

CHAPTER 4
Rendering

4.1 Introduction

In the last two chapters we saw how we can model and animate three-dimensional objects. We saw a wire frame representation of the objects in both cases. Wire frame images are easy to produce but they don't do a good job of depicting objects as they appear in real life. After the 3D models have been generated, we would like to define them to have a more photorealistic *look*. How can we achieve visual realism in our images?

The quest for visual realism is multifaceted. First we need to identify which surfaces are visible from the current viewpoint. The surfaces not visible are removed from the scene, a process called hidden surface removal. Once visible surfaces have been identified, we could simply assign a single color to each surface and paint it accordingly. However, in real life, we see that surfaces are shaded differently due to different amounts of light incident upon them. This is evident when an object is illuminated in a dark room with a flashlight. Surfaces facing the flashlight are bright, while surfaces facing away are dark. If the room had no lights, we would only see darkness. The next step then is to define the lights available in the scene and assign the visible surfaces with surface properties like color. The computer uses these variables in a shading algorithm to generate realistic shaded images.

The step in computer graphics in which we assign special attributes to a model (such as surface color), add lights to our computer graphics world, and create a shaded image is called rendering. Rendering is a critical process in computer graphics and can be quite time-consuming depending on the complexity of the image and the shading algorithm used. The shading algorithm does not precisely simulate the behavior of light and surfaces in the real world, but approximates actual conditions. Different shading algorithms yield different results. Some work well in one given situation while others work well in other situations. The design of the algorithm is a trade-off between precision and computing cost. For example, a typical image in the film *Toy Story* which required high precision took anywhere between an hour to 24 hours to render on a powerful workstation! Most of us would like to see our renders come up immediately instead of waiting a day to see the results. In this book, we employ some simple shading algorithms in order to see our rendered images within a few seconds. The quality of the image is not that which one would see in a movie, but it's good enough for our purposes.

Let us explore the different phases involved in rendering.

4.2 Hidden Surface Removal

When we talk about surfaces in this book, we mean the polygonal shell that defines the shape of the object. For our purposes, a surface is composed of one or more adjacent polygons.

In the modeling section, we saw that a wire frame representation made objects look transparent. We learned how to eliminate the *back facing* polygonal surfaces of an object. But what about those surfaces that are obscured by other objects in front of it? Fig. 4.1 shows a wire frame model of a sphere being partly obscured by a cylinder that lies in front of it. In real life, not only would the back surfaces of the sphere

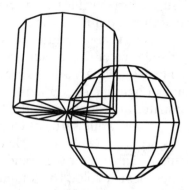

Fig. 4.1: A cylinder partially obstructing a sphere.

and cylinder be hidden from our view, but the portion of the sphere behind the cylinder would not be visible to us either. The first step in rendering is to remove such *hidden* surfaces. This process is called hidden surface removal. Back face culling is part of the hidden surface removal procedure.

One of the simplest technique to accomplish hidden surface removal is known as *z-buffering*. In this process, a buffer (called a z-buffer) is used to keep track of which polygon is the closest to the eye at any given pixel. The closest polygon at a given pixel finally gets drawn onto the screen, while those further away get eliminated as they are obscured by this polygon at that pixel. Note that this process can be used to automatically eliminate the back facing polygons also. The process we employed to do back face culling in the modeling chapter no longer needs to be used with such a technique.

Surface removal is but one factor in generating a realistic image. Every surface needs to be shaded in accordance with the lights illuminating it and its surface material properties. Let us look into how we define lights in our CG world.

4.3 Lights

Imagine entering a dark room with no lights in it. If the dark room is perfectly insulated from any kind of light, you will most likely not be able to see anything in the room even though there are objects in the room. In a similar manner, if we do not define light sources in our CG world, we shall not be able to *see* the objects that exist in the scene. Imagine now that we switch on a bulb in the room. Immediately we shall start seeing the objects in the room. How brightly we see them will depend on the wattage of the bulb. The color of the light will also affect the look of objects. They may appear reddish under a red light, or eerily blue under a blue light. The bulb is the light source enabling us to *see* the material objects in the room.

In CG, as in real life, we need to define lights to illuminate a scene. A light source has a color and an intensity to describe its properties. We saw in the first chapter how color can be represented as a triplet of RGB values, where the red, green, and blue components have a value between 0 and 255. We shall follow this convention to describe the color of lights. The intensity of light can have a value between 0 and 1. The intensity is simply a scale factor for the color of the light. The actual color of the light is the product of its color and its intensity. It is easier to think in terms of intensity when dimming a light or brightening it. An intensity of 0 means the light is off, and an intensity of 1 means the light is at its brightest. Intensity can be thought of as a knob that controls the brightness of the light.

The light sources we use in CG can be of different kinds. The most commonly used types of lights are

- Ambient Light
- Distant Light
- Point Light
- Spot Light

Some lights need a position and a direction in addition to the color to be completely defined; others need a direction, and still others don't need either.

For the purpose of our book, we only use ambient and distant lights, but we describe all the lights in detail for the sake of completeness.

Ambient Light

In some situations, we do not worry about where the light is originating from or towards which direction it is pointing. We define this light to be shining uniformly in all directions. This kind of light that has no direct source, and is prevalent uniformly in a given area, is called ambient light.

An ambient light distributes light uniformly through space in all directions. It throws light on all surfaces no matter how they are oriented. If a scene has only ambient light, then all objects will appear to be shaded with no variation across their surface. This can lead to a dull two-dimensional look. Fig. 4.2 shows a sphere that is illuminated by only ambient light. Notice how flat the sphere looks as it is uniformly lit in all directions.

Fig. 4.2: Ambient Lighting on a sphere.

In most three-dimensional computer graphics scenes, it is common to add a small amount of ambient light to prevent any surface from being completely unilluminated (and hence appearing black).

Distant Light

The light from a distant light flows uniformly in space from one direction. As a result, surfaces with the same orientation receive the same amount of light independent of location. Surfaces of different orientations are illuminated differently. Those facing toward the light source

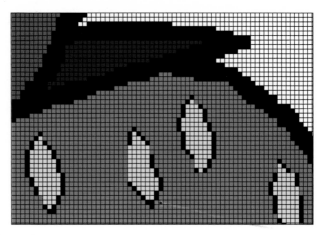

Color Plate 1: Pixel view of a strawberry image.

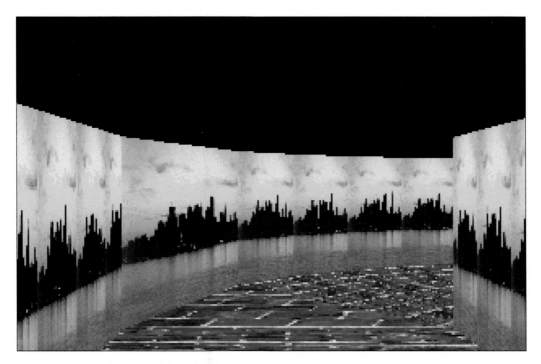

Color Plate 2: Gallery model with tiling textures.

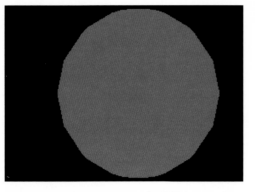 

Color Plate 3: Spheres with different values of K_a, K_d, K_s.

Color Plate 4: Utah teapot with different values of K_a, K_d, K_s.

 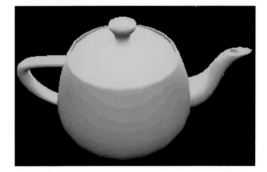

Color Plate 5:
Composite
sequence for
Snowman.

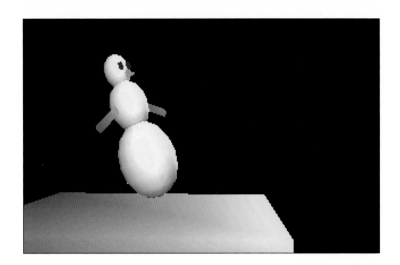

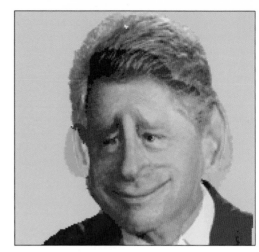

Color Plate 6: Morph sequence (Bill Clinton to Ross Perot).

appear brightest, while those facing away are completely unilluminated by this light source within the scene.

The best example of a distant light is the sun. As viewed from the earth, its rays are essentially parallel and position makes no practical difference on the intensity of the light. In Fig. 4.3, a distant light is shown lighting up a sphere. The distant light is represented by directed lines to indicate from which direction the light is coming. Notice how the side of the sphere facing away from the light's direction is completely black.

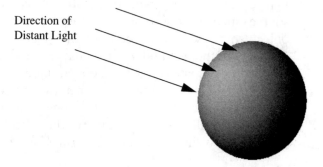

Fig. 4.3: Distant Light on a sphere.

Point Light

The point light source distributes light through space from a single point. It shines evenly in all directions. The intensity of the light is usually defined to fall off with the square of the distance from the light to the surface. You could, however, define it to have no fall off at all. This light needs a position and a direction to be completely defined. A bulb is a good example of a point light.

Spot Light

A spot light, like a point light, is a light source with both position and direction. It simulates a cone of light emitted from one point to another point. The intensity of this light is defined to fall off exponentially with angle from the center of the cone. This kind of light is what you often see in theater lighting and is often used to direct the viewer's eyes to certain parts in the scene.

As mentioned before, we do not provide the point light and the spot-light in our software. You will find ambient and distant lights to be more than adequate to produce great looking results.

4.4 Surface Materials

Different objects tend to have different surface properties. We perceive these properties all the time in real life. A shiny apple has a smooth and shiny surface; a piece of paper, on the other hand, tends to look dull and matte. How do our eyes perceive the color and material properties of an object?

Every object reflects light that hits it. In Fig. 4.4 we show a beam of light hitting a spherical object. This lightbeam is called the incident light, as it is incident on the object. It may have originated from a light source (such as the sun) or it may have been reflected from some other object (this is when you would see an object being reflected in another object). For the remainder of the chapter we shall assume that all incident light originates either as a distant light or an ambient light. This incident light is then reflected. When this reflected light hits our eye, we *see* the object. The color of the reflected light will be the color that we perceive. Also depending on how the reflected light gets scattered, we perceive a shiny and smooth surface or a matte dull material.

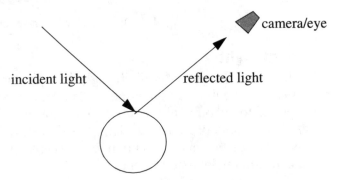

Fig. 4.4: Light being reflected off the surface of a sphere.

Our task in rendering is to determine the outgoing color and direction of light that hits objects and finally hits our CG camera. To do this we need to know the color, direction, and type of light sources in the world. We also need to know the material properties of the objects in order to determine how light is reflected from the surface of an object.

The most essential property of a surface is its color. We define the color of a material by an RGB triplet, similar to the way we describe the color of light. The color of the material will dictate a large part of how the reflected light looks. We shall assume the color of the surface in question to be given by the RGB triplet, (R_s, G_s, B_s), and the intensity and color of the incident light to be I_i and (R_i, G_i, B_i), respectively. This means the perceived color of the light is actually $(I_i*R_i, I_i*G_i, I_i*B_i)$, which are values between 0 and 255. We can normalize these values to be between 0 and 1 by dividing each component by 256. You can think of these normalized components as the intensity of each color of the light.

There are three different shading models that we use to calculate the reflected light. They are

- ambient reflection

- diffuse reflection

- specular reflection

The simplest model is the illumination resulting from ambient light. We call this reflection ambient reflection.

Ambient Reflection

Ambient reflection is the name given to surface reflection due to ambient light sources. The intensity of the reflection is independent of the surface location and orientation. We define a coefficient of ambient reflection, K_a, for a material. Similarly, we define K_d as the coefficient for diffuse reflection and K_s as a coefficient for specular reflection, which we shall talk about in a moment.

K_a has a range of values between 0 and 1. If K_a is 0, then the surface will not have any ambient reflection. This means that any ambient light in the scene will have no effect on the illumination of the object. If K_a is 1 then the object will reflect all the ambient light with which it is being illuminated. Most surfaces have a K_a of around 0.1 to 0.2. Sometimes you may want to make a surface look flat and bright, like maybe a model of the sun. You would then define the surface K_a to have a high value like 0.7 or 0.8 or maybe even 1.

If the object surface has a color (R_s,G_s,B_s), and if the incident light has an intensity I_i and a color (R_i,G_i,B_i), then the reflected light due to ambient reflection will have an intensity of 1 and a color (R_{ra},G_{ra},B_{ra}) given by

$$R_{ra} = (I_i*R_i)/256*R_s*K_a$$
$$G_{ra} = (I_i*G_i)/256*G_s*K_a$$
$$B_{ra} = (I_i*B_i)/256*B_s*K_a$$

Note that we use the color intensity of the incident light described earlier to multiply into the color of the surface and the coefficient of ambient reflection, K_a. This ensures that the final result, is a color with a value between 0 and 255.

■
**Mini
Project**

- Start up the render program by clicking the render.exe icon. Refer to Appendix C for more information.

- The interface will be similar to the model and animation software. You will notice that the display window has suddenly become smaller. The reason for this will be apparent in the next chapter when we combine our rendered images with other graphic images.

- Create a sphere with a radius of about 40 units. You can do this by choosing the Create option under the Object menu and clicking the Sphere option. Set the radius of this sphere to 40 units.

- Set the complexity to about 3 by choosing the Complexity option under the Draw menu. This will make the sphere's surface appear smoother. You can also set back face culling on if you wish.

- Our first task is to define a material with a color and a coefficient of ambient reflection, K_a. To do this click on the **Materials** menu and choose the **Create** option. This will bring up a form that will let you edit in different material properties. Let us try to define a material that has a yellowish appearance to it.

- In the **Color** frame, you have sliders to control the Red, Green, and Blue values of the color of the material we are defining. Try sliding them around. The box to the right of the Color sliders will give you some idea of what color you are setting. This, however, may not be the exact color you will see in your rendered image, but it will be close enough to give you some idea of what color you are defining. Set the color to a dull yellow by setting the Red slider to 100, the Green slider to 100, and let the Blue slider remain at 0.

- In the **Coefficients** frame, set K_a to be 0.3.

- The Material Name frame lets you edit in a name for the material. Clear out the default name and type in **GoldMat**. This sets the name of the material to be GoldMat.

- Ignore the other options for the moment. Click OK when you are done.

- You have now created a material with a color (100,100,0) having a coefficient of ambient reflection $K_a = 0.3$.

- We now need to assign the sphere to have this material. To do this, click the Materials menu and choose the **Assign Material** option. This will bring up a form with a list of materials and a list of objects to which you can assign the material to. At the moment only the GoldMat material will show up in the materials list, and the sphere will show up in the object list. Pick the sphere object and the GoldMat material in the lists by clicking on them (and hence highlighting it on the form). Click OK to assign the GoldMat material to the sphere.

- Using this technique, you can define different material to assign to your objects. At the moment the software allows you to define up to eight unique materials.

- Let us define a light source to light up our sphere. We only need to define an ambient light source, as only ambient lights will be reflected at the moment. To define a light, click the Lights menu and choose the Create option. This will bring up a form that will let you edit in properties for a light source.

- Set the Color of the light to have an RGB value of around (100,100,100). This is a greyish-white light. The Light Type will already be Ambient Light, as the default. Ignore all the other columns for the moment. Set the Lightname to be **AmbLt** and click OK to create this grey ambient light.

- Using this method you can define up to eight unique lights. Only the first light that you define is allowed to be an ambient light. All the other seven lights MUST be distant lights. The first light can be either an ambient or a distant light.

- Now click the Rendering Mode menu. This menu allows you to change the mode of the display. Currently we have a wire frame display. Click on the **Flat Shading** option (and we shall explain what this kind of shading is later) to change the mode. This will render an image on the screen for you.

- What do you see? Do you see a flat yellowish circular object? The shading has no variation across its surface causing it to look rather two-dimensional and flat, as ambient reflection illuminates all polygons to be a constant shade regardless of orientation.

- Try changing the coefficient of reflection. To do this, you will need to pick the GoldMat material. Under the material menu, choose the Pick option and select GoldMat from the form. (At the moment this will be the only material in the form.)

- Now pick Modify from under the Materials menu. This will bring up the Materials form again, and will allow you to modify the appropriate entries. Slide K_a up and down and see what results you get in the display screen. Reset K_a to 0.3 and click OK when you are done.

- You can also try modifying the light intensity and color. To do this, you will need to pick the AmbLt light. Choose the Pick option under the Lights menu and pick AmbLt from the form.

- Now pick the Modify option under the Lights menu. This will bring up a form to let you modify the light. Play around with the colors of the light to see what effect it has on the final rendered image. Notice that changing the blue component of the light will have no effect on the image. Why is that? Recall that the blue component reflected is a direct multiple of the blue color of the surface. Since we have defined the surface to have no blue component, the reflected light also has a zero blue component. Also try changing the intensity of the light. At an intensity of 0, the image rendered will be completely black, as the light is now switched off.

- Once you are done experimenting, reset the light color to (100,100,100) and an intensity of 1.

- At this point, you can save the materials and lights you have set up by choosing the **Save Lights and Materials** option under the File menu. This will bring up a file requestor prompting you for the name of the file in which to save the attributes. Save the file in **c:\3dcg\Models** with the filename **goldamb**. This file gets appended by a suffix of mtl. This file will hence be saved out as **goldamb.mtl**.

Diffuse Reflection

Although objects illuminated by ambient light are lit in proportion to the ambient intensity, they are uniformly illuminated across their surfaces. Most objects we see in the world are not shaded uniformly. Their surface has some variation in shading across it. This kind of illumination can be simulated by diffuse reflection. Distant lights (and for that matter, point lights and spot lights) have two components, the diffuse and the specular component. Diffuse reflection is the surface reflection of the diffuse component of distant lights.

Distant lights have a definite direction in which their light rays propagate. Most surfaces will exhibit some amount of diffuse reflection. Such a reflection causes the surface to appear to be shaded differently depending on the orientation of the surface and direction of the incident light. The viewing direction has no effect on the brightness of the surface. That is to say, even if you change the camera position, it will have no effect on the illumination of the surface caused due to diffuse reflection. In Fig. 4.4, we show a polygonal surface with normal N being hit with incident light I. Diffuse reflection is scaled by the cosine of the angle between N and I (t, as shown in Fig. 4.5). This means that light that hits the surface straight on (at t=0) will cause the brightest amount of light to be reflected diffusely. Light that hits at grazing angles (t>=90) will cause little or no light to be reflected.

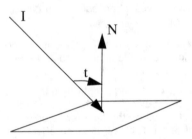

Fig. 4.5: Diffuse reflection depends on the angle t.

We define the coefficient for diffuse reflection for a surface to be K_d, whose value is between 0 and 1. At a K_d of 1, the surface reflects all the diffuse components of light hitting it, and at a K_d of 0 it reflects none. If we define the diffusely reflected light to have the color (R_{rd}, G_{rd}, B_{rd}), they can be defined as

$R_{rd} = (I_i * R_i)/256 * K_d * R_s * \cos(t)$

$G_{rd} = (I_i * G_i)/256 * K_d * G_s * \cos(t)$

$B_{rd} = (I_i * B_i)/256 * K_d * B_s * \cos(t)$

In Fig 4.6 we show a series of spheres with different K_d values. The surface itself has a grey color (100,100,100) and is being lit by a distant grey light. The value of K_d varies from object to object. A dark-grained wood object that doesn't respond much to light would have a low K_d like around 0.2, whereas a light-colored plastic object that reflects a

$$K_d = 0.4 \qquad\qquad K_d = 0.6 \qquad\qquad K_d = 0.8$$

Fig. 4.6: Surface at different values of K_d.

lot of light would have a high K_d, like maybe 0.8. Of course the final choice of K_d used for any object in the scene depends finally on what looks *realistic* and appealing to your eye.

Mini Project

- Assuming you still have the render program running, let us modify the GoldMat material to include a K_d component. We shall also add a distant light to this scene.

- Pick the GoldMat material and choose the Modify option under the Materials Menu.

- Set K_d to be about 0.5. Make sure you have reset the color to have an RGB of (100,100,0) and K_a to 0.3 from last session.

- Create a new light. Let it have a grey color of (100,100,100). This will be defined as a distant light by default. Now distant lights have a direction in which they are shining. The AngZ and AngY sliders will let you change the directions of the light source. AngZ and AngY define the angle between the direction of the distant light and the world axes, respectively. Initially the light is pointed straight down in the negative Y direction.

- Set the Light Name to be **DistLt** and click on OK to create it.

- Notice the sphere in the Display window. You should now see your image change in appearance. It will be illuminated from the top. The illumination, however, is not uniform: the lower polygons at the base of the sphere get no diffuse light at all. The polygons on the top of the surface of the sphere that are oriented towards the direction of the light are the most brightly lit.

Let us take a break from the project for a moment to understand how we can control the direction in which the distant light is pointing. The direction of the distant light can be visualized as a vector which points in the direction of the incident light rays. Initially the light is defined to be directed straight down along the negative Y axis, as shown in Fig. 4.7a.

Changing AngZ causes the direction of the light to be rotated by an angle of AngZ about the Z axis. Visually, you can imagine our vector being rotated about the Z axis to be oriented at an angle of AngZ with respect to the Y axis in the X-Y plane, as shown in Fig. 4.7b

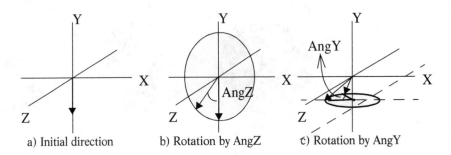

a) Initial direction   b) Rotation by AngZ   c) Rotation by AngY

Fig. 4.7: The Light Direction Vector.

Changing AngY causes the direction of light to be rotated about the Y axis by an angle of AngY. Visually, you can imagine our vector to now be rotated about the Y axis to achieve its final orientation, its tip describing the arc of a circle on a plane parallel to the X-Z plane. The angle subtended by the arc is AngY, as shown in Fig. 4.7c. It may be obvious from the figure that at AngZ=0, this circle will have a radius of zero, and hence changing AngY will have no effect on the light direction. The orientation of the vector gives us the final direction of the light.

Remember that rotation is not commutative! We always define AngZ to occur before AngY, to ensure predetermined results for any rotation. Continuing on with the exercise.

Mini Project

- Let us modify the DistLt light by changing its direction by changing the values of the AngZ and AngY sliders. Set AngZ to be -60. What result do you see? Is it what you expect?

- If you bring up the Schematic View window (by choosing **Schematic View** option under the Show menu) you will see icons for distant lights that will give you an idea of where the light is pointed. The distant light is represented by a circle icon with a line pointing in the direction of the light. Note that since this is only the top view, you will not be able to judge what angle the light ray is oriented with respect to the Y axis by looking at the display. Only the direction in the XZ plane will be obvious by looking at the display. This is good enough to give you a sense of where the light is pointing.

- As you change AngY, notice how the light icon swings around in the schematic view. Do you understand how the direction of the light is affecting your image?

- Set AngY to be -44 to define the light source to be in front, above and to the left of the sphere.

- Try modifying the K_a and K_d of the GoldMat material and observe the image in the display. At a K_a of 0 you will notice that some parts of the sphere are completely unilluminated, as they are not facing the direction of the distant light. Also experiment with changing the color of the material to observe what results you get. Do not exit from the program yet. If you do, please save your materials and lights in a file.

- Try experimenting by adding more lights to the scene. How is your image get affected?

Specular Reflection

Specular reflection can be observed on any shiny surface. Take an apple in your hand and illuminate it with a bright light like a torch. Do you notice a highlight on the surface of the apple? The highlight is caused by specular reflection. The light reflected from the rest of the apple is diffuse reflection. Note that at the highlight, the apple appears to be white in color (or whatever the color of the torch is) and not the color of the apple. Now move your head (or the viewing position) and notice how the highlight also moves. It does so because specular reflection depends on the direction from which the surface is being viewed.

In general, specular reflection depends primarily on

- the orientation of the surface (or the normal of the surface)

- the direction of incident light

- The direction from which the surface is being viewed (or the camera viewpoint)

We shall not concern ourselves with the math behind how specular reflection works, as it is beyond the scope of this book. We instead define a function specular(), which depends on the above three factors. We define a coefficient for specular reflection, K_s, which takes a value between 0 and 1. Since the specular reflection has a different color than the color of the object, we define a specular color for objects. This specular color is a function of the material of the surface. If this specular color is given by (R_{sc}, G_{sc}, B_{sc}), the specular reflection component of the reflected light is then given by (R_{rs}, G_{rs}, B_{rs}), defined as

$$R_{rs} = (I_i * R_i)/256 * K_s * R_{sc} * \text{specular}()$$

$$G_{rs} = (I_i * G_i)/256 * K_s * G_{sc} * \text{specular}()$$

$$B_{rs} = (I_i * B_i)/256 * K_s * B_{sc} * \text{specular}()$$

Throughout this book, we assume that the specular color for surfaces is white; that is it has an RGB value of (255,255,255). Fig. 4.8 shows a series of images of a sphere at different levels of Ks.

The specular function is a property of the material. For those curious, it is defined as $\cos^n(\alpha)$, where $\cos(\alpha)$ is defined as the dot product of the direction of reflection and the viewpoint direction, shown in Fig. 4.9. n is the materials's specular reflection exponent. Please refer to the references for more details about specular reflection.

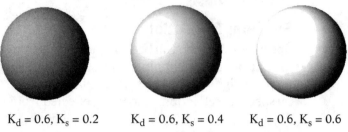

$K_d = 0.6, K_s = 0.2$ $K_d = 0.6, K_s = 0.4$ $K_d = 0.6, K_s = 0.6$

Fig. 4.8: A sphere at different values of Ks.

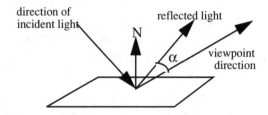

direction of
incident light

reflected light

N

α viewpoint
direction

Fig. 4.9: Specular reflection.

**Mini
Project**

• We assume you still have render running at this stage. (If you quit from the program, please restart render and load up the materials file you had saved the GoldMat material in. Create a sphere again and assign the GoldMat material to this sphere. Change the Rendering mode to be Flat.)

• Modify the GoldMat material to have a K_s of 0.2. Do you notice any change in the rendered image?

• At this point, change the rendering mode to be Gouraud (we shall explain these modes in a moment). This will give you a smoother and more accurate rendering of your model. Plate 3 has a color representaion of the sphere rendered with the ambient, diffuse, and specular components turned on.

• Try sliding K_s up and down and observe what you see. Do you see the highlight in the sphere? Try changing the direction of the distant light and see what results you get.

• You can also try rotating the camera around the sphere to see how the highlight is affected. Use the Transforms option under the Camera menu to change the camera viewing position.

• Now that you understand the concepts of ambient, diffuse reflections, let us try a more interesting model. Change the rendering mode back to Wire frame. Load in the teapot model by choosing the **Open Model** option under the File menu. Choose the **teapot.mdl** file from the **c:\3dcg\Models** folder.

• Click on the 'Assign Material' option under the Materials menu and assign the GoldMat material to the Teapot.

• Now change the rendering mode back to Gourad. You will see your teapot rendered in gold! Plate 4 shows the teapot rendered with diferent coefficients of reflection.

- Try modifying the GoldMat material to see what results you get. Do they match with the results you got with the sphere?

- If you wish to exit the render program at this stage, please save the materials and light settings you have created in a file.

Specular reflections are important to the perceived realism of objects. Few surfaces are so diffuse that the don't have a degree of shininess to them. We usually assign objects to have a small amount of specular in most CG scenes. A shiny object like an apple can have a specular like 0.6 or 0.7, whereas a dull matte object like a paper may have a K_s of 0.1.

4.5 Shading Algorithm

In the last few sections we saw how we can find the color at every pixel location in our CG scene by certain shading calculations. Performing the shading calculation amounts to computing the color of every pixel on every surface in the model and varying the colors to account for uneven lighting effects. Performing so many calculations can take forever to compute. Instead there are some techniques where shading calculations are limited to some key points on the surface and the rest of the points are interpolated. We shall look into two such shading techniques: Flat shading and Gouraud shading. In Flat shading, we calculate the color of each polygon based on the normal vector of the polygon. The entire polygon is then given this color. This results in a very faceted-looking model, which you may have noticed in our previous exercises. Gouraud shading does shading one better by calculating the color of every vertex of every polygon and blending the polygonal color.

By themselves neither of the shading techniques account for some of the more complex real-life lighting effects such as shadows and reflections. Advanced lighting and shading algorithms exist that allow such effects but take comparatively longer to compute each image. Let us look into the details of Flat Shading.

Flat Shading

Flat shading adds realism to models in a scene by correlating brightness of each surface to the angle of the light incident upon it. This shading method is good for polygonal surfaces. The procedure is explained below:

1.   For each polygonal surface the normal vector is computed.

2.   Next the angle, t, between each normal vector and the direction of incident light from an imaginary light source, is computed. The camera

viewing direction V is known, as it is supplied by the user. Knowing these factors, we can calculate the ambient, diffuse, and specular brightness of each polygon. These components are computed once for a polygon and the brightness does not vary across an individual polygon. A polygon with a normal vector parallel to the light source has the highest color intensity.

3. This procedure is repeated for each polygon and when hidden surfaces (surfaces obscured by other polygons) are removed, the image produced is a flat shaded image.

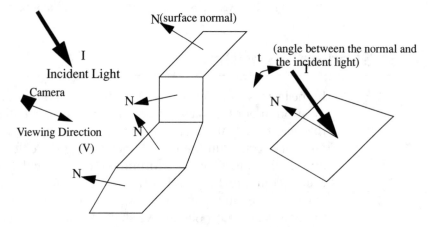

Fig. 4.10: Calculating Flat Shading.

In the flat shading scheme, the polygon surfaces that make up the surface of the object are all uniformly lighted. Flat shading is usually assigned to surfaces on objects that are small in dimensions. Due to the small size of objects,there is no need to vary brightness across the surface of the object.

The flat shading method is fastest to render, as the intensity across the polygon surface stays constant and each surface pixel has the same brightness. Flat shading is used to preview rendering effects when speed of rendering is critical. Flat shading is also utilized in situations where there are only a few colors available in the color system. We already saw this kind of shading in our previous examples, so we shall not run any exercise at this point.

For making surfaces look smoother, rounder, and less faceted Gouraud shading is used.

Gouraud Shading

Gouraud (pronounce goo-Row) shading is used when smooth shading effects are needed. Unlike flat shading, Gouraud shading varies intensity of light and computes the color of individual pixels across a polygonal surface. The shading procedure is explained below:

1. Compute the normal vector for each vertex of a polygon surface. The vertex polygon normal is found by averaging normals of all polygons that meet at that vertex.

2. The camera viewing direction V is known, as it is supplied by the user. Knowing these factors, we can calculate the ambient, diffuse, and specular brightness at each vertex of the polygon.

3. Colors on the edges and in the interior of the polygon are interpolated from colors of the corners. This varies the brightness at different points on the polygon surface and adds a degree of realism to the shading.

4. This process is repeated for all the polygon surfaces and yields a Gouraud shaded object.

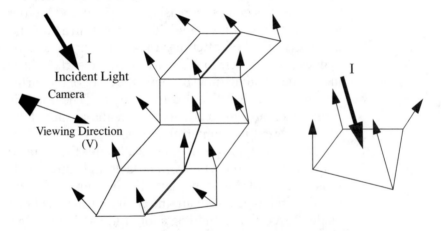

Fig. 4.11: Calculating Gouraud Shading.

Gouraud shading yields smoother renders for curved or spherical objects. The lines between polygon edges blend together to produce a smooth look. However, the tradeoff is that Gouraud shading takes longer to complete and also rounded edges still have slightly angular silhouettes.

Even more complicated rendering techniques exist, each making the scene more realistic, but trading off on computation time. You may have heard of such techniques like ray tracing and radiosity. These techniques are too advanced for the scope of this book. Interested readers should refer to the bibliography for more information on these topics.

■ **Mini**
  Project

- If you have the render program running from last session, change the rendering mode to Gouraud. If not, please re-start the render program. Create a sphere, and load up the materials file you have saved the GoldMat material Assign the sphere to have the GoldMat material, and then change the rendering model to be Gouraud.

- Pick the DistLt light and change AngY and AngZ to view what effects the lighting has on gouraud shaded polygons.

- Notice how in Gouraud shading the sphere appears to be shaded more uniformly compared to flat shading. Try changing the AngZ of the DISTANT light, Toggle between the Flat and the Gouraud rendering modes and compare the two results.

4.6   Texture Mapping

In the last few sections we saw how we can assign surface characteristics to a modeled object in order to add a sense of realism. However, in some situations it becomes increasingly difficult and tedious to model and assign characteristics to individual objects in a complicated scene. An alternative method often used is to superimpose a completely independent two-dimensional image over the surface of the object. This means the 2D image now becomes the color of the surface of the object. This technique is called texture mapping. The image itself is referred to as the texture, and it is mapped onto the surface of the desired object using different mapping techniques.

Texture mapping can dramatically alter the surface characteristics of an object. It adds vitality to models and can provide great looking visual cues to simple models. For example, if we texture map a woodgrain image to a chair, the chair's surface will now look like it is made of wood grain, adding a lot of life and meaning to the modeled chair.

We have defined textures to be true replicas of the image. This means the texture gives the object the same color as the original image. Our textures do not respond to light, hence dimming the lights will have no effect on the look of the polygon. In more advanced software the texture will modulate with the light to create a more realistic look.

You can use any BMP image as a texture map. You can create your own images using your own favorite paint program. The only restriction on the image is that it be rectangular, with its dimensions in pixels a factor of 2, like 128, 256, 512 etc. This means the image you choose must have a size (in pixels) such as [128 by 256], [256 by 512], [512 by 128], etc.

There are different ways of mapping a texture onto an object. Advanced software will let you specify how the texture should be mapped onto the surface of any given object. We describe a few common approaches for mapping textures to the basic models. The ones we describe in this book are

- Flat mapping

- Wrapping

- Combination and Specialized mappings

- Tiling

For some objects, we use a combination of two or more methods to map the final texture.

The simplest method is called Flat Mapping.

Flat mapping

Flat mapping is normally used for flat surfaces like polygons. Flat projection works much like a slide projector mapping the image to the *surface* of a wall. We use flat mapping to map polygons. Given a four-sided polygon, Fig. 4.12 shows how the image is mapped onto the polygon.

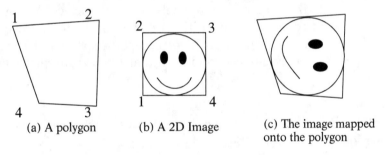

(a) A polygon   (b) A 2D Image   (c) The image mapped onto the polygon

Fig. 4.12: Mapping a Polygon.

The vertex points of the polygon each get assigned to a corner of the image, as shown. The image is stretched to fit the four sides of the polygon and then projected onto its surface.

Mini Project

- Restart the rendering program if you have exited from it. If not, go ahead and delete all the models from the last session. Note that the old materials and lights will not be deleted when you do this.

- Create a polygon with vertices (0,0,0), (0,60,0), (40,60,0), and (60,0,0), in that order. If you have forgotten how to do this, you will need to click the Create option under the Object menu and choose the Polygon option. Then slide the XYZ sliders to set each vertex.

- Under the Materials menu, select the Create option to create a new material. Check the box labelled Texture. This will bring up an Open File requestor asking you to pick an image to be the texture. We have provided some textures for you in the **c:\3dcg\Textures** directory. If you wish to get your own BMP images, make sure you resize them to the size we discussed before. For right now, pick the **mandrill.bmp** texture supplied by us. The label box under Texture will display the name of the image you have picked as the texture. Name this material as mandrill and click OK.

- Assign the polygon to have this material.

- Change the rendering mode to be either Flat or Gouraud. Both will yield the same result, as the lighting calculation is not taking place in our texture mapping. Notice how the polygon surface is now the mandrill image? (For those of you who don't know, a mandrill is a type of monkey.)

Flat mapping works fine for flat surfaces. For curved surfaces, however, we start running into distortions. Imagine the slide projector projecting an image onto a curved surface. The image would distort at the curved edges. To avoid this distortion, we use another technique of mapping images. This technique is called wrapping.

Wrapping

Wrapping works like a shrink wrap. The image follows the curve of the object. A sphere is an ideal candidate for carrying out wrapping. You can imagine the image to be a piece of cloth that is then fit around the sphere to completely cover its surface. You will get seams at the edges where the different edges of the image meet.

Internally, each polygon of the sphere is getting assigned a section of the texture map such that the texture map warps itself smoothly around the sphere.

**Mini
Project**

- Change the Rendering mode back to WireFrame.

- Create a sphere model.

- Assign the sphere to have the mandrill material created in the last session.

- Change the Rendering mode back to Gouraud.

- You will now see the sphere has a new look! The mandrill image has been wrapped around its surface! Rotate the sphere by choosing the Transform option under the Objects menu to observe how the texture has been wrapped around it.

Combination and Specialized Mappings

Because of the mathematical formulae used, flat mapping works best for flat surfaces and wrap projection works best for curved surfaces. For objects like a cylinder or a cone, we use a combination of the two methods to wrap the texture around it, as shown in Fig. 4.13. For these objects, we have two images(three in the case of the cylinder) being mapped on, one using the wrapping technique, the other by a flat map.

The box primitive has flat mapping on all its faces. The only difference while mapping onto a box is the way the texture is finally applied to it. The entire image is not mapped onto each face. Instead, the entire image is first wrapped around the vertical faces of the cube. The tops and bottoms are then folded onto the top and bottom face of the cube.

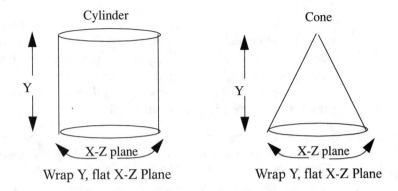

Fig. 4.13: Map types for cylinder and cone.

The rest of the image is discarded. Fig. 4.14 shows how this splits the image before each section is applied using a flat mapping to each face of the cube. Sections 1, 2, 3, and 4 are mapped to the vertical faces of

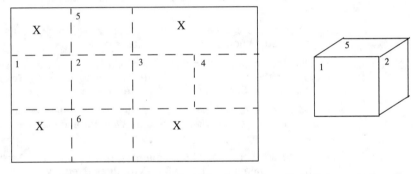

Fig.4.14: Flat mapping on a cube. The image is split before mapping to eac face of the cube.

the cube. 5 is applied to the top and 4 is applied to the bottom of the cube. Its almost like wrapping a Christmas gift box with wrapping paper. The sections marked X are discarded. This technique is one of the ways we could have wrapped the box. Although there are a number of ways wrapping can actually occur on the box, this technique is the most popular. As an exercise, create a cube and try mapping the mandrill texture onto its surface. Observe the results and see if you can identify how the image was broken down.

The disk simply has the image flatly mapped onto it. The image is slightly cropped along the perimeter of the disk so as to not distort the image. Try creating a disk and mapping a texture onto it.

Tiling

All our objects are essentially composed of component polygons. In tiling, each component polygon of the object gets a flat map of the image, as described earlier. Since each polygon gets the same image map, the result looks quite like a tile pattern, hence the name.

Let us explore the tiling map by loading an animation we created in the last chapter, the *gallery* animation.

Mini Project

- Reset the render mode to be WireFrame and set the World option on under the Draw menu. You can turn on back face culling if you desire.

- Load up the gallery animation that you saved from the last chapter in **c:\3dcg\Models** directory. Since you are loading an animation, a warning will come up telling you that all the current models will be deleted once the animation has been loaded. Go ahead and click OK.

- The previously created models will be deleted and you will now see the gallery from the last chapter.

- In this case we need to define two materials, one for the side walls, and one for the ground.

- Let us create a material having the San Francisco skyline as a tiled map. Click the Create option under the Materials menu.

- Click the box next to **Texture**. The Open File requestor will come up prompting you for the name of the texture file. Go to **c:\3dcg\Textures** and choose **sfo.bmp**, the San Francisco skyline texture we have provided for you.

- Set the Tile Texture button on by clicking the small box to the left of the **Tile** option. Set the Material name to be **WallMat** and click OK to create this material.

- Create another material having a tiling texture with a brick texture. This will be the **brick.bmp** file in the **c:\3dcg\Textures** directory.

- Set the Tile Texture on.

- Rename this material as **GroundMat** and click OK.

- Assign the WallMat material to the object called **Walls** using the Assign Materials option.

- Assign the GroundMat material to the **Ground** object.

- Change the Rendering mode to Gouraud. Do you see how the textures are tiling across on each polygon, defining the walls and the ground? Plate 2 has a color image of what you should be seeing on your screen.

- Make the animation by choosing the Make Animation option under the Animation options menu. This will make the animation for you. Play the animation back and observe your gallery animation come to life!

- Go ahead and exit from the render program.

At this stage let us try to consolidate all we have learned by applying it to our Snowy animation.

4.7 The Snowy Animation

Now that we have learned the basics in rendering, let us try to apply our knowledge to the snowman animation we created earlier.

Here's an outline of what we need to do to make this animation come alive.

- Load the animation
- Identify and create materials for the different objects in the scene, and assign them.
- Identify and create lights that enhance the mood of the scene

In the next section we shall learn how to composite this animation with a colorful background to really make it feel real.

Mini Project

- Bring up the snowman animation and see how we can add more color to this scene.
- Restart your render software by clicking the Render icon.
- Open the snoman animation by clicking on the Open Animation under the File menu and selecting the snowman animation file saved from the previous chapter. This will load up the models and the final snowman animation created. Remember to turn on the Draw World option so you can see all the models. You can turn on back face culling if you desire.
- Let us first identify which materials are needed for this scene. We need a snow material to define the snowman's body. Snow is white with a slightly blue glaze look to it. Under the Materials menu, click the Create option. Set the color to be whitish blue; and RGB of about (100,100,110). Set K_a to be about 0.2, K_d to about 0.5, and K_s be 0.2. Rename this material as **SnowMat**.
- We need a carrot-like material to define the nose. Define a new material having a color of (210,61,0) to define an orange color. Since carrots are not very shiny, we shall let K_s for this material be 0. Set K_a to be 0.2, and K_d to be 0.6. Name this material **CarrotMat**.
- Next we need a grey nonspecular material to define the eyes. Define a material with a color of (55,55,55), having a K_a of 0.2 and K_d of 0.5. Name this material **EyeMat**.
- We assume the hands are made of tree branches and hence should be brown in color. Wood is not shiny at all, so we shall not define this material to have any specular coefficient. Define a material called **WoodMat** having a color of (99,56,42), a K_a of 0.2, and K_d of 0.4
- Although this may seem a bit tedious to you, bear with us. You are building a library of materials you can pick from for any new animations!
- Now we need to assign the materials to the right objects.
- Assign the SnowMat material to the Head, Tummy, Base, and Plateau object (which shows up as Box_9 if you haven't renamed it). Choose the Assign option under the Materials menu and click on these objects to highlight them. If you press the Ctrl key while picking the objects, you can select more than one entry at a time. Pick the SnowMat material and then click APPLY to assign it to the

objects selected.

Note: The APPLY will assign your material to the selected objects and will also keep the 'Assign Material' form open for you to assign other materials. This is useful when we have many materials in a scene which we need to asisgn to objects. Clicking on OK also assigns the material, but will close the 'Assign Materials' form.

- Similarly, assign the EyeMat material to the eyes (LeftEye and RightEye), the CarrotMat material to the nose. When assigning the WoodMat material to the hands (LeftHand and RightHand). Click on OK to assign the material and also close the form. We are done assigning materials to all the objects in the scene.

- We now need to define some lights. When defining lights you should think about where the light sources may be coming from to make the scene more realistic. In this scene we may assume that there is a sun shining behind Snowy somewhere far to the right of the screen. This is where our distant light should be coming from. It should also be yellowish-white to match the color of the sun. The ambient light we can define to be a cool blue, the ambient light from the blue sky. We shall also define a third white light coming from below the snowman. Although this light has no logical source, you can imagine it to be the light reflected off the snow to light up the snowman from beneath. It also helps to soften the black bottoms of the snowman.

- Create an ambient light having a color of (200,200,215). Set its intensity down to about 0.5. Call this light **SkyLight**.

- Create a distant light having a color of (130,130,50), a pale yellow sunlight. Set its direction to be AngX = -54 and Angy = 45. This will define a light coming from the screen right and from behind Snowy. Name this light **SunLight**.

- Create a second distant light having a color of (120,120,120) and an intensity of 0.4. Set its direction to have an AngX of -64 and AngY of -100. This defines the direction of this light to be pointed upwards. Set the light name to be **SnowLight**.

- Change the Rendering mode to be Gouraud shading. What do you see? Has the scene completely transformed into something more realistic? The background is still black. In the next chapter we shall see how to assign a cool backdrop to this animation. Of course, you could have also just created a flat polygon behind the snowman and texture mapped it with a texture to create the backdrop.

- When we create animations, the program stores the images in a temporary file to which it refers when playing the animation. Let us save the rendered snowman animation frames to have a unique file name. In the next chapter we shall see how we can enhance our final images to add even more realism.

- Under the Animation options menu, choose the **Save Shot As** option and fill in **snowman** in the text box. This will save all our frames in the **c:\3dcg\Images\Anim** directory as snowman##.bmp, where ## is the frame number going from 01 to 60.

- Make the Animation.

- Wait for the animation to be completely built. When you play back the animation, it will be in color! Play back your animation and watch Snowy in his new real-life settings! The individual frames will be saved in the Animation directory for those curious to look at them.

- You will notice one peculiarity in your animation. In the last few frames when Snowy looks down, his nose seems to disappear inside his tummy! Recall that we had created this animation in Wireframe mode. This is a problem animating in wire frame. A wire frame model gives you a good sense of the shape of objects, but only when you finally render the models can you observe all the finer details. We encourage you to modify the snowman animation to take care of this little problem. You will need to modify the key frame when the head looks down so that the intersection does not take place.

- You can save the materials and lights you just created by choosing the **Save Light and Materials** option under the File menu. Name the file snowman to be consistent with our snowman trend. (If you look into this directory this will be saved out as snowman.mtl). Remember that if you ever load it up you will still need to assign the materials to the respective objects. We have provided you with a backup **snowman.mtl** file in the **c:\3dcg\Models\Examples** directory in case you need it.

- Change the rendering mode to be Flat. What changes do you see in the appearance of the scene? Can you now see the difference between the two rendering techniques?

4.8 Summary

In this chapter we discussed the concepts involved in rendering an image. We learned about different kinds of lights employed by graphics users and also the different kinds of reflections that materials exhibit.

We learned how shading algorithms like Flat and Gouraud shading are used to calculate the color value of select points in our scene to speed up rendering. Finally, we applied our knowledge of rendering to render the gallery animation and the snowman animation from the last chapter.

You are now an expert in three-dimensional graphics! If you have understood the concepts of the last four chapters, well, then any new software or book in this area will be easy for you to grasp. In the next chapter we take you to one final step that most computer production houses use to clean up and add special effects to the final images rendered.

CHAPTER 5
Postproduction

5.1 Introduction

The use of computer graphics in movies and commercials is very widespread. In the past few chapters, we saw how to model, animate, and finally render our 3D shots into a sequence of images. These images, when played back at a certain speed, gave the illusion of motion.

The images you created were rendered through software that we provided for you. We saw in the introduction that there are many other ways of inputting images into the computer. One could scan in photographs or use a sequence of images from a video. You can get images by downloading them from the Internet. You could even paint your own images using a paint package like Pbrush (which comes as part of the Microsoft Windows package). We also saw that these images can be stored in a variety of formats, which we discuss in Appendix A.

Once you have all the necessary sequences of images, they need to be processed to achieve the final finished animation. This final stage in the production sequence is often called Postproduction. In the Postproduction phase all the final rendered images are edited, arranged, and organized. In a typical production a team of editors, animators, and producers would sit together during this stage and decide when and where scene changes take place, how different images are combined together, and how images need to be touched up. The team may also employ image manipulation techniques in this stage to achieve special effects. Today effects like morphing and compositing are routinely used in productions. These Postproduction techniques are used to create dazzling effects on the silver screen and also in TV commercials. We will look at some special effects tricks used by animation whizzes in this chapter. Using the software provided, you will be able to create your own spectacular special effects.

5.2 The Images

In the final stage of production, rendered images are given their final form. The format we use in this book is the Windows bitmap (*bmp*) file format developed by Microsoft. This file format is very popular on the personal computer and is supported by most programs that run on the Windows platform. It is also the default format for the Windows Pbrush program, so you will not need any special paint program to paint or view your bmp images. Let us briefly discuss the technical details of the bmp file format.

Bitmap File Format

The bitmap file format is the most common graphics format used on the Windows platform. The format comes from Microsoft Corp. and is the standard format for all Windows applications. Although *bmp* is based on Windows internal data structure, it is supported by many non-PC and non-Windows applications, too.

The motivation behind creating the *bmp* format was to store raster image data in a format that is independent of the color scheme used on any particular hardware system. The color schemes supported in BMP are monochrome, color-lookup table, and also RGB colors (i.e., 24-bit color). The data in BMP files is stored sequentially in the form of a 2D array of pixel values. The data is stored in binary format and sometimes the data may be compressed using RLE compression (Run-Length-Encoded).

Table 1 shows the detailed format of a bmp file with the bitmap file header, bitmap information, and bitmap data sections. The bitmap data for pixels is stored by rows, left to right within each row. The rows are stored bottom to top; the origin of the bitmap is the lower-left corner. The bits of each pixel are packed into bytes and each scan line is padded with zeros, if needed, to be aligned with a 32-bit boundary.

Table 1:

Byte #	Data	Details
1-2	File Type	Must be ASCII text "BM"
3-6	Size of the file	In double words (32-bit integers)

Table 1:

Byte #	Data	Details
7-10	Reserved for future use	Must be zero
11-14	Byte offset to bmp data	offset from bmp file header to start of file
Byte#	Data	Details
1-4	Number of bytes in header	Currently 40 bytes
5-8	Width of bitmap	in pixels
9-12	Height of bitmap	in pixels
13-14	Number of color planes	Must be set to 1
15-16	Number of bits per pixel	1 or 4 or 8 or 24; default 24; determines palette size
17-20	Type of compression	0: no compression; 1: run length(8 bits per pixel); 2: run length(4 bits per pixel)
21-24	Size of Image	in bytes
25-28	Horizontal resolution	In pixels/meter
29-32	Vertical resolution	In pixels/meter
33-36	Number of color indexes used by bitmap	Zero indicates all colors are important
37-40	Number of color indexes important to display bitmap	Zero indicates all colors are important
41	Blue color value	Beginning color palette (entry 0)-blue value

Table 1:

Byte #	Data	Details
42	Green color value	Beginning color palette (entry 0)- green value
43	Red color value	Beginning color palette (entry 0)- red value
44	Reserved for future use	Must be zero
....	Remaining color palette entries	4 bytes per color palette entry, the number of entries is based on bits per pixel value above. Colors are listed in order of importance

Table 1: The bitmap file format

Let us now proceed to learn about image manipulation, enhancement, and touchups, and how to use tricks to create transitions and special effects on our graphics images.

5.3 Image Enhancement

Once the final sequence of images is ready, it is usually desirable to touch them up before printing them or sending them to film. Such touchups can be easily achieved by a variety of paint programs available on the market. Depending on the sophistication of the paint package, you can perform a variety of touchups and image enhancements on the image. PhotoShop put out by Adobe is one of the most sophisticated paint packages available.

Microsoft provides a free basic paint package called Pbrush with Windows 95. This package can be used to load in a *bmp* image and to paint over undesired portions of the image, or paint in a new item into the image. You can try painting on top of some of snowman images saved from the last chapter. You can try painting a blue background instead of the black background that the snowman images have at the moment. We do not go into the details of how to use Pbrush, interested readers should be able to refer to the online help documentation provided by Microsoft.

Transitions

In most animations there are times when you want to switch from one shot to the other. A shot is a sequence of images that have been captured from a given camera position and direction. A different shot occurs when we change the camera location or camera orientation in a scene. Normally the switch from one shot to another is done by combining the shots together. In general, this kind of change is called a transition. Usually a transition is a cut, which means simply switching to the new shot on the very next frame without any effect. This is what we did in the snowman animation in the animation chapter. In the snowman sequence we see the snowman bouncing up and down and then cut to the next shot, where he suddenly realizes he is off the cliff. In some situations a more dramatic transition is called for, such as slowly fading in one image while fading out another. This kind of transition is called a dissolve. A dissolve forms the basis of most complicated transitions. Let us look into the process of dissolving one image frame into another.

Image dissolving is a process of adding of two digital images. We studied linear interpolation techniques in the animation chapter. We use the same equation to linearly interpolate the colors of each pixel location in the first image with the corresponding pixel value in the second image to compute the resultant image pixel. We refer to the first image as the source image and the second image as the target image (or simply source and target). Look at Fig. 5.1, where each image is represented as a matrix of pixel values.

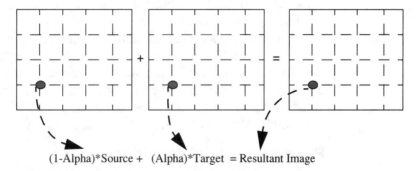

(1-Alpha)*Source + (Alpha)*Target = Resultant Image

Fig. 5.1: Dissolving two images by adding pixel values.

When we refer to pixel values, we are actually referring to the color values at each pixel location. We studied about the RGB color system and how we can define the color of each pixel by a 24-bit value for a true-color system. For a true-color system, dissolving is a simple matter of linear combination of the colors at each pixel location from the source and destination images. The process of linear interpolation of pixel colors is also called alpha blending.

The equation set up for each pixel of the resultant image is

pixel_intensity (of resultant image)=(1-alpha)*source pixel+(alpha)*target pixel.

At alpha=0, pixel_intensity of resultant image=source pixel, that is, the resultant image is equal to the source image.

At alpha=1, pixel_intensity of resultant image=target pixel, the resultant image is equal to the target image.

So as we change alpha from 0 to 1.0, the dissolved image transitions from source to target. If we changed alpha slowly over time, we would get a smooth fadeout of the source image and a fade in of the target image. If we have two shots, s1 and s2, we would blend the last frame of s1 with the first frame of s2 to get a smooth transition from one shot to the next.

From the above discussion, alpha blending or dissolving is a straightforward, linear, additive process. This is certainly so in the case of computer systems that support true colors or 24-bit colors. But most personal computers do not support a 24-bit color system. For systems using color maps, the color maps of the source, target, and resultant images may not be the same. Typically PCs and the software available for them (including ours) support 256 colors. When we blend two images (with their set of unique 256 colors or 256 color maps), some of the colors we get by blending the source and target colors may not be in our resultant (dissolved) 256-color set. How do we handle these colors that do not belong to the 256-color set?

There are several algorithms available that compress the resultant *blended* colors to a 256 color table that may be different from the one used by the source and/or target images. We have picked one called the "*Popularity Cut Off*" method. As the name suggests, in this method we begin by picking the 256 most popular colors. For the remaining colors, we try to find the closest matching colors in the selected 256-color table. Here is an outline of the algorithm we have adopted.

Popularity Cut Off Algorithm

Let us look at a simple example to understand the *popularity cut off* algorithm.

Given: An image with a resolution of 4 x 3 pixels. The image has a color resolution of 3 bits, one bit each for the red, green, and blue color components. The image originally has eight colors, represented by the numbers 1 through 8, as shown in Fig. 5.2. In this figure each

2	4	7	5
2	4	6	1
1	7	2	8

Fig. 5.2: Sample image with 4 x 3 resolution.

square grid represents a pixel and the number inside each pixel is the color value of that pixel.

Task: To compress the color space of this image to allow only four colors using the popularity algorithm.

Procedure:

We begin by making a table of the frequency of each color in the image by reading in each pixel color. This gives us the frequency table of each color (as shown in Table 2). For example, the color 2 occurs three times in the image and so its frequency is listed as 3 in Table 2.

Table 2: Frequency table

Red	Green	Blue	Color	Frequency of color in source Image
0	0	0	1	2
0	0	1	2	3
0	1	0	3	0
1	0	0	4	2
0	1	1	5	1
1	0	1	6	1
1	1	0	7	2
1	1	1	8	1

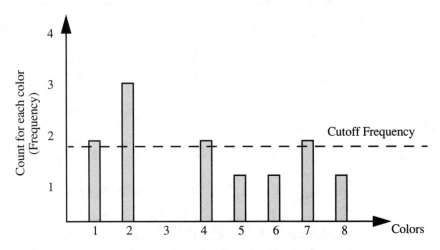

Fig. 5.3: The Frequency chart of colors in source image.

If we take color 1 to represent the minimum value of the colors, then color 1 represents black color (no color), and color 8 represents white (the brightest color). Can you tell what colors 2 through 6 represent?

We now start building a count of the topmost frequencies till we hit the maximum number of colors allowed in the compressed color space (in our example maximum colors is four).

As seen in the frequency chart in Fig. 5.3, the color 2 occurs the most times. This color is picked up and assigned to our color table. There are three colors that occur with a frequency of two; these are colors 1, 2, and 7. If we pick up these three colors, our color table will have four colors: 1, 2, 4, and 7. Since this is the maximum number of colors allowed, no more colors are picked up at this stage.

We need to assign all the pixels in our image to colors in our four color table consisting of colors 1, 2, 4, and 7. Any pixel having a color already in the table retains its original color. The remaining colors are assigned to map to one of the selected colors. We would like to map each color to a color that matches it closest in hue and saturation. This will ensure that the resultant image does not vary too much in look from the original image. To find the closest color, we make use of the distance formula to find color distances.

Given any two colors, i and j, the distance of color i from color j is given by

Distance (color i from color j)=sqrt[(Ri-Rj)**2+(Gi-Gj)**2+(Bi-Bj**2)]

where Ri, Bi and Gi represents the Red, Green and Blue component of color i, and Rj, Bj, and Gj represent the red, green and blue components of color j.

For example, take color 5. The distance of color 5 from the four colors in the selected color table is

Distance (Color1, Color5)=sqrt[((0-0)*(0-0)+(1-0)*(1-0)+(1-0)*(1-0)]= sqrt(1+1)=sqrt(2)=1.414

Distance(Color2,Color5)=1

Distance(Color4,Color5)=sqrt(3)=1.7

Distance(Color7,Color5)=sqrt(2)=1.4

From this calculation, color 5 is closest to color 2, and hence pixels with color 5 will now be represented by color 2. We apply these calculations to every pixel in the source image till all colors have been assigned to a color in the color table. The new re-mapped image is shown in Fig. 5.4b.

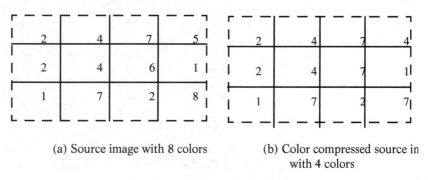

| | (a) Source image with 8 colors | | | | | | (b) Color compressed source in with 4 colors |

(a) Source image with 8 colors   (b) Color compressed source in with 4 colors

Fig. 5.4: The Re-Mapped Image.

The popularity cutoff algorithm works well on images with a small number of colors. For larger number of colors, the resultant image tends to look washed out. Also if the images have small highlights, they may not get colored properly in the dissolved image.

Several other methods of color compression are commonly used in CG. The *Median cut* algorithm makes use of the median of the popular colors and thereby ensures that small details are not left out. If a fast color compression algorithm is needed, the *Fixed Color palette* method is used. In the *fixed color palette* system, the 256 colors in the palette are predetermined and fixed. All the resultant colors are then mapped to these fixed 256 colors.

Another popular method used for displaying more colors is called *dithering*. In *dithering,* pixels of differing colors next to each other are shown in such a way that it produces an illusion of having more colors. Common examples of dithering are the newspaper and a TV screen.

If you look closely at the print on a newspaper you can clearly see how the black-and-white color dots are dithered to give an illusion of different levels of gray.

Well, you must be waiting to see a dissolve for yourself. Let us try dissolving an image of a strawberry into that of a sunflower. At this point we assume you have installed the software supplement that accompanied this book. If not, please do so and refer to the installation instructions.

Mini Project

• Start up the Postproduction program by clicking on the pprod.exe icon. Refer to Appendix C if you have forgotten how to start up the program.

• This will bring up an interface, shown in Fig. 5.5. This interface allows you to select from a suite of tools you can use for Postproduction.

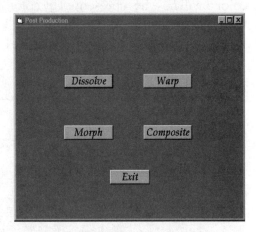

Fig. 5.5: Postproduction Main Form

• Choose the Dissolve tool by clicking the button marked **Dissolve**. The interface needed for the dissolve process will come up.

• From this interface, click on the File menu, click on Open and select the **Source Image** option. This will bring up the dialog box for selecting your source bmp image. Choose the **strawberry.bmp** file from the **c:\3dcg\Images** directory.

• This will load the strawberry.bmp image in the picture window in the Dissolve form.

• Again select the File menu, click on Open, and this time select the **Target Image** option. This will bring up the dialog box for selecting your target bmp image. Choose **sunflower.bmp** file from the **c:\3dcg\Images** directory.

• This will load the sunflower.bmp image in the picture window in the Main form.

• You are now ready to dissolve the source and target images. From the Run menu, select the **Execute** option. (A shortcut to execute this command is hitting the **F5** key.) This will invoke the dissolve program to create your dissolved image.

- Wait for the **Done with Dissolving** message dialog box to come up. This will indicate that the dissolve process has completed. Click OK.

- The Control Panel to play frames will appear in the lower-right corner of the dissolve form. This panel is the same from the last two chapters.

- At this point we have only one dissolved image. From our discussion on alpha blending, you may have guessed that an alpha of 0.5 was used to generate this one intermediate image which is halfway between the Source and the Target images. You can click the **Play_Frame** button to go through each frame in our dissolve.

- You can also click the Play button and the program will loop through the source, dissolved, and target images once. To make this loop play continuously, select the Loop button in the playback frame.

- At the moment, we have three images in our dissolve animation, with one transition image.

- To create more intermediate dissolved images, select the Options menu button and click the **Set Transition Frame Count** option. This will bring up a scroll bar to let you select how many intermediate images you want your dissolve to create. The maximum number of transition frames allowed is 30. However, each image will make use of significant disk space, so we recommend you start with a small number of transitions (say around 5–10). Also make sure your computer has enough disk space to store these images.

- Move the scroll bar to set the transition frame count to 4 and then the OK button to set these changes.

- Now run the dissolve program again by pressing the F5 button.

- After the Done dialog box comes up, click OK and play the series of dissolved images using the Play Frame buttons. The program has now created four intermediate dissolved images for you. Can you tell what values of alpha are being used to generate the intermediate frames?

- Now that we have introduced you to the dissolve tool, you can use your own bmp images to create some cool dissolve effects. While using your own bmp images, please make sure your images are:

  - Bitmap images with 8 bit or less color resolution with no RLE compression (Note: In order to find out if bmp images are RLE-compressed, load the images in Pbrush in Windows. If the images load up in Pbrush, then they are not RLE-compressed)

  - Of size less than 400 x 300 resolution

  - The source and target images should be of the same size.

- If you want to save your dissolved images, you will need to use the following option before you execute the dissolve program. Select the File menu option and click the **Save Output As** option. This will bring up a Save dialog box. Type only the prefix of the output filename; for example, mydissolve. The program will store the intermediate dissolved files in the **C:\3dcg\Images\Pprod** directory as <filename>##.bmp file, where # is an digit (e.g., mydissolve01.bmp, mydissolve02.bmp, etc.) depending on the number of transition frames you select. Note by default, the program saves output files with the prefix **tmp**. After selecting the output filename, you will have to execute the dissolve program to save the output images. Once they are ready, you can load these images up into Pbrush and examine them or modify them as you please.

- You can exit from the Dissolve program by choosing the exit option under the File menu. This will bring you back to the Postproduction main form. To exit from this form, simply click the button marked Exit. You can leave the program running if you wish to continue with the next section.

5.4 Special Effects

Special Effects like morphing and warping are routinely used by movie makers and also by advertising agencies in creating commercials. The process involves manipulating the pixels of the images to achieve eye-catching results. Applications of special effects do not stop here. computer graphics effects are also used by NASA, the weather service, educators, artists, magazine publishers, and the list goes on. With such widespread popularity, it seems worthwhile to learn some tricks of the graphics trade.

Digital Image Warping

In the warping procedure we distort the outlines of a given source image to create a warped image. Let us carry out a simple exercise to understand the concept of warping.

Imagine you are holding a large elastic sheet with a printed image of *smiley* on its face, as shown in Fig. 5.6a. Let us assume that the rubber sheet is pinned down along its four edges. Now let us stretch or pinch the rubber sheet at arbitrary locations. The smiley face on the rubber sheet will get distorted and change depending on where we stretch or pinch the rubber sheet. In Fig. 5.6b, the rubber sheet has been stretched out in the horizontal direction by pulling it on the sides, as shown.

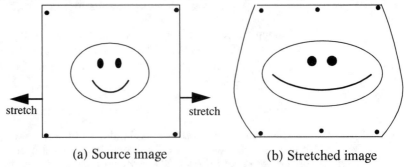

(a) Source image   (b) Stretched image

Fig. 5.6: Rubber sheet analogy of warping an image.

The rubber sheet analogy can be applied to digital warping. Warping is essentially a set of geometric transformations we apply to an image to produce the stretching or pinching effect. The warping procedure incorporates some important features, as seen by our rubber sheet experiment. They are listed below:

- We can define a warp by pinching or stretching the image at a few points or locations. We shall call these locations control points/ lines that define the warp.
- The effects of warping seem to be local to the control locations. For example, when we stretch the smile on the face of the rubber sheet, the eyes do not get distorted to the same extent.
- The image as a whole always remains connected.

These properties are useful in developing a computer warping tool. The warp algorithm rotates, stretches, squashes, and transforms pixels of the image depending on how the control locations are manipulated. So the first step in beginning the warping procedure would be to define a set of control locations that can be manipulated. We make use of control lines (lines connected by two points). An original set of control lines (called the source control lines) are first defined on the image. These control lines are then modified in various ways to define the target control lines for the image. In the warp procedure, the source control lines are mapped onto the target control lines. The task of the warp program is to manipulate the image pixels by moving them in the same way as their local source control lines are being moved to attain the final orientation of the target control lines.

The warping algorithm translates into a range of mathematical operations to be performed on all pixels of the image. The warp control lines define the basis for these equations. Some math details of the warp routine are explained below.

Math Stuff

In our warping algorithm we make use of lines to stretch, squash, or rotate the image. For purposes of clarity, we will assume we have one source control line (XY) on the image to be warped to the target control line (X'Y'), as shown in Fig. 5.7. The source control line is modified to a new location and orientation to define the target control line. This movement of the control line causes all the pixel locations on the source image to move. The critical parameters that define the warping of each source pixel S to target pixel S' are:

- The perpendicular distance, d, of the pixel from the control line
- The distance f of this perpendicular from X, the end point of the control line

As shown in Fig. 5.7, the source pixel S is moved to its final location S'. S' is defined such that the perpendicular distance of S' from the target control line X'Y' is the same as that of S from XY. Also, the fractional distance of this perpendicular from X' with respect to the target line X'Y' is the same as that of f with respect to the control line XY. In math terms,

SM=S'M'=d and,

f/XY = f'/X'Y' (= X'M'/X'Y')

If we were to extend this to two or more control lines, the influence of each of the control lines can be weighed in by incorporating the distance of pixel points from each of the control lines and adding the result to achieve the final pixel location, as shown in Fig. 5.8.

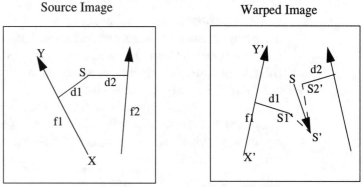

Fig. 5.8: Two or more source control lines define weight-averaged pixel location.

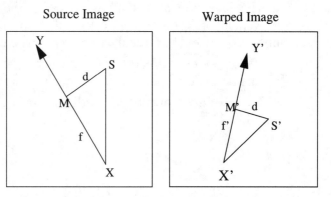

Fig. 5.7: The parameters for warping.

If we have exhausted you with the math talk, let's take a break. Let us look at warping some images and creating our own special effects. Here we go ...

Mini Project

- Load the Postproduction program by starting the pprod.exe program. Select the Warp tool. This will bring up the interface needed for the warp process.

- Click the File menu option and select the **Source Image** option. This will bring up the dialog box for selecting your source bmp image. Choose the **diet.bmp** file from the **c:\3dcg\Images** directory.

- This will load the diet.bmp image in the Picture window in the Warp form.

- After you select the source image, the DrawLines frame will also appear on the right side of the form. Click the DrawSrcLines option and draw three horizontal lines on the bmp image, as shown in Fig. 5.9a. The central dot on each line marks the center of that line and we will use it later to edit lines.

- Next click the DrawDstLines option. The source image with source control lines will be displayed in the picture window. In the DrawDstLines mode, you can edit the source control lines to define the target control lines needed to define the warp. The control lines can be edited in two ways:

    1. Grab the ends of the source lines by clicking the mouse at either of the end points and moving the lines to desired lengths and positions (the control lines have rubber-banding effect).

    2. Click the central red dot of a control line and move the entire control line

to a new location (without changing its length).

(a) Diet.bmp image with source control lines.

(b) Diet.bmp image with target control lines.

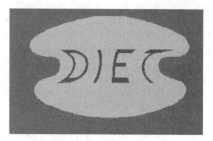

(c) Warped Diet.bmp image

Fig. 5.9: Warping an image.

Note that the destination control lines can only be defined after the source control lines have been drawn.

- For the diet.bmp file, edit only the middle control line by grabbing its end points and reducing its length, as shown in Fig .5.9b.

- Next click the Options menu and set the transition frame count to 3. The warp algorithm only needs a source image. The target image is the final warped source image. Setting the number of transition frame counts to 3 will generate four warped images, the final one being the target image. So we shall get four frames in our warp sequence.

- You are now ready to warp. Press the **F5** button or select Execute under the Run menu option.

- Wait for the **Done Warping** message dialog box to come up.

- The Play Control Panel will come up, allowing you to play back the warped images. Click on the Play button and watch the words **Diet** actually squeeze in to form an hour glass shape! The pixels around the control lines are getting warped in different ways depending on the proximity to the control line, leading to such a warping effect. Fig. 5.9c shows a warped image of the diet.bmp file.

- Try different settings of control lines. Choosing **Draw Control Lines** under the Options menu will allow you to reset the source and destination control lines.

- Try warping different images we have provided in the **C:\3dcg\Images** directory. Note that the program has a maximum of 10 control lines. Experiment with different speeds of play_back. Enterprising readers can create their own bmp images using Pbrush, to generate their own custom warps! Remember that the image size should not exceed 400x300 pixels.

- As with dissolve, you can save your transition frames by choosing the **Save Output As** command under the File menu and typing in a unique filename. The transition frames will get saved in **c:\3dcg\Images\Pprod\<filename>##.bmp**.

- You can exit from the program when you are ready.

Morphing

Morphing is an acronym for "metamorphosing." This term comes from biology. It means the transformation of a creature from its larval (or infant) form into its adult form. Some common examples are a tadpole metamorphosing into a frog or a caterpillar changing into a butterfly. In each of these morphs, the key features of the infant develop into key features in the adult form.

In computer graphics the definition of morphing is used in a broader sense. Morphing in computer graphics is the transformation of any object (living or not) into another object with similar features. This may remind you about the morphs you have seen in movies and TV advertisements. (it is hard to miss this phenomenon these days!) In the movie *Interview with a Vampire* actor Tom Cruise morphs into a vampire; the TV commercial for Lexus shows a tiger morphing into a Lexus car. Well if you have not seen any of these morphs, don't be disappointed. We are going to introduce you to some fascinating tools that will help you create your own special effects. Also, if you are adventurous, you can create your own animation clips employing the special effects learned in this chapter.

Morphing is a combination of the two processes we have learned so far, warping and dissolving. In the morphing process we select two images, the source image and the target image. As we morph the source to target, we go through the warp and dissolve processes. In the morphing sequence at each stage, the location of pixels on the intermediate image is found using the warp algorithm, and its color is set using the dissolve algorithm. Also, at each step the color map of the morphed image is found by employing the popularity algorithm. Let's see how this is implemented by morphing some caricatures of famous characters.

Consider morphing Billy into Ross as shown in Fig. 5.10. This sequence is also shown in color in Plate 6. Note how in the morphing sequence shown, Billy is being warped to fill the outline of Ross and Ross is being warped backward to fill the outline of Billy. Both warp sequences are blended together with Billy fading out and Ross fading in — a dissolve. Let us look further into the workings of the morph.

As in the warping algorithm, we first need to assign control lines on the source image (Billy in our example). The key to making a clean

Billy.bmp
(source image)

Morphed Images

Ross.bmp
(target image)

Fig. 5.10: Morphing sequence showing Billy morphing into Ross.

morph is to make sure that the control lines on the source image are placed on matching features on the target image. This will ensure that key features of the source image *grow* into the target image features, producing a smooth morphing phenomenon. So for our current example, we will need certain key features of Billy and Ross to align. We match control lines (shown in Fig. 5.11) on the following key features of the two characters:

- eyes
- nose
- ears
- hair line
- jaw bone.

You can understand the morphing sequence better after trying it out, so let's play with morphing.

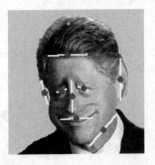

(a) Source image with control lines   (b) Target Image with control lir

Fig. 5.11: Images with control lines.

Mini Project

- Load the Postproduction program by starting the pprod.exe program and click the Morph button.

- Click the File menu option and select the Source Image option. This will bring up the dialog box for selecting your source bmp image. Choose the **billy.bmp** file from the **c:\3dcg\Images** directory. This will load the billy.bmp image in the Picture window in the Morph form.

- After you have selected your source images, the option buttons for drawing control lines would have become visible. We need to select the control lines on the source and target images to ensure that key features of both images will align during the morph process. Click on the DrawSrcLines option. Draw the control lines on Billy, as shown in Fig. 5.11a.

- Next load the **ross.bmp** file as your target image.

- The DrawDstLines option should now be marked. If it is not, click this option. This will let you edit the control lines to define the target lines. To edit the control lines, you can grab any of the ends of the lines and stretch them (creating a rubberbanding effect), or you can click the red blob at the center of each line and move the line while holding the mouse button down. Move the control lines to align the features of Ross, as shown in Fig. 5.11b.

- Set the number of transition frames to 4.

- You are now ready to morph. Click the Run option menu and select execute or press the **F5** button and wait for the **Done Morphing** message dialog box to come up. Note that morphing is a time-intensive process, as it has to warp and dissolve each pixel point on both the source and target images. Depending on the size of your image, the number of transition frames chosen, and also the number of control lines used, a morph process can take several minutes to complete.

- Click the OK in the dialog box. The Play Control Panel will come up. Click the Play icon and the display window will show you the morphed images. We show some of the intermediate frames in Fig. 5.11.

- Well, now you are ready to be a morphing expert. So we will leave you to handle the next morph sequence on your own. Here's some data and tips for your morphing assignment. Choose **cpillar.bmp** and **butterfly.bmp** as your source and target images. This will load up a caterpillar and a butterfly, as shown in Fig. 5.12. The images also show suggested control lines to see a smooth morph sequence. Of course you can add more control lines to achieve different effects. Remember that you cannot choose more than 10 control lines and also the more the control lines and the more the number of transitions, the longer will be your morph time. Try out this morph!

Fig. 5.12: Assignment for Morphing a caterpillar into a butterfly.

Have fun morphing other images!!

Compositing

Compositing is another common tool used in Postproduction. In compositing two, or more different images from different sources are combined into one image in such a way that an illusion of time and space is created. When we look at a composited image, it appears as though the images happened at the same time and place and were also recorded or captured together.

In the rendering chapter, we rendered a snowman bouncing up and down against a black background. Suppose now you wanted to make this snowman look as if it were flying in front of San Francisco? If you could combine these two images in the right way, you could achieve this effect without having to actually wait till you saw a snowman bouncing in front of San Francisco! The process of combining independent images to make a new combined image is called *Compositing*.

A good analogy to a composite image is a collage. A collage is an assembly of image fragments or materials from different sources. To create a collage, one would use scissors to cut the different fragments and then glue them together on a sheet of paper. To composite an image in computer graphics, we employ what is called a mask or matte to get different portions from different sources. The masks help us cut out the relevant fragments of each image and then glue them together on the composite image.

Why is compositing so popular? Well, compositing can eliminate expensive production costs for complicated images. Compositing can merge together images from different sources like live video shots, 3D rendered images, or 2D hand-drawn images. The flexibility to combine sources can empower an artist to generate physically impossible and surreal visual effects. Care must be taken to ensure that the lighting of the different images match up, so that the composite image looks believable. A series of composited images can be used to create fascinating animation sequences. In this section we will make use of a simple and flexible technique to composite images and also create our own composited animation.

Compositing obviously needs two images, the foreground image and a background image against which it will be composited. One of the easiest ways to achieve the composite is to specify one of the colors in the foreground image to be *transparent*. The pixels of the background image show through the *transparent* pixels of the foreground image. The pixels that are not transparent, of course remain un-altered. Since the pixels may have some amount of variation in their color, you can specify an error range for the 'transparent color'. This transparent color is also called matte color. Any color which falls within the specified error range of the matte color is assumed to be transparent. As in the case of dissolved images, since we deal with only 256 color sets, the resultant composited image may have more than 256 colors. In compositing we also make use of the popularity algorithm to find the 256 most popular colors in the composite image and assign the popular colors as the image color map.

The compositing software provided with the book can composite foreground images over a selected background image. You first load in the background image and then load the foreground. You can pick a matte color on the foreground image. All pixels on the foreground image with the matte color will become transparent. The composite software will replace these transparent pixels with pixels from the background image. This process will then yield us our glued (composite) image. Our software constrains the two images to be of the same size in pixels. A background image with a different size than the foreground image will yield unpredictable results in the composite image. In the rendering chapter, we generated all images at a size of 320x200 pixels. This is a fairly common size, and you can easily find images of this size from the Internet. Fig. 5.13 shows one frame from our snowman animation as the foreground image composited with the background image of the San Francisco skyline.

Enough of the theory. Let's go on to the Mini Project and create our own composite animation.

Mini Project

- Load the postproduction program by starting the pprod.exe program. In the Programs menu, select the Composite option. This will bring up the interface needed for the Composite process.

- Click the File Menu option and select the BackgroundImage option. This will bring up the dialog box for selecting your background bmp image. Go to the **C:\3dcg\Images** directory and choose **sfo1.bmp** as your background image. This image is the skyline of San Francisco. Remember to go to the correct directory when choosing the image. In the last chapter, we had loaded a San Francisco skyline from the **C:\3dcg\Textures** directory and the size of this image was 256 X 128 in accordance with the texture image file requirements. The image we want has a size of 320 X 200, the same size as the image of the snowman animation.

- Next click the ForegroundImage option in the File menu and choose **snowman01.bmp** as your foreground image.

- We need to choose a transparent or matte color in the foreground image to carry out our compositing. Click your mouse on any location to pick the black color on the image (where the snowman is not drawn). This color will show up in the Matte Color box.

- Run compositing by pressing the F5 button and wait for the **Done** dialog box.

- At the end of the composite process, the composite image will be loaded in the picture window. Note how the transparent pixels in the foreground image have been replaced by pixels from the background image.

- Now we can create an animation using composite images. Select the Set Shot length option under the Options menu and select the number of frames to be 30, to load up the next 30 frames of the snowman animation. Thirty is the maximum number of frames we allow to be composited at one time, due to time and hard disk space issues.

- The composite program will now load a sequence of 30 foreground images starting from snowman01 (snowman01, snowman02, snowman03, ...,snowman30) and composite each of them with the background image. Press the F5 button to composite the images.

- Fig. 5.14 shows a composite frame from this snowman animation. This image is shown in color in Plate 5.

- When the playframe comes up, click the Play button and enjoy your composite animation sequence of the snowman bouncing in front of San Francisco! (Note you can also click the ForegroundImage option and watch the foreground images being played as an animation, or change the matte color).

- You can save the composited animation frames by choosing the **Save Output As** under the File menu and typing in the file name. The composite images will be saved in **c:\3dcg\Images\Pprod\<filename>##.bmp**, where ## will be the frame number. You will have to press **F5** again to create these images.

- You can load up snowman31 as the new foreground image and composite the remaining 30 frames of the animation, if you have enough disk space.

BackGround Image

ForeGround Image

Composite Image

Fig. 5.14 Compositing images

- You may notice a slight discrepancy when the shot changes at frame 41. At this point, the camera angle changes, causing the perspective on the foreground image to shift. For a realistic effect, we would like the background image to shift in perspective also. However, this does not happen, as the background image was a static image recorded earlier. To make the second shot look realistic, you need a background image recorded from the new camera perspective, and composite the last 20 frames of the animation with this new background image. This matching up of the background image to account for camera moves is often called **camera match move**. We have provided a new image called sfo2.bmp for this purpose. We leave the compositing as an exercise to the reader.

- You can experiment with your own bmp images and composite them with any background image. Remember to define a matte color for the foreground images and also an error range for the matte color if needed. To animate a composite sequence, your foreground images should be named **<filename>##.bmp**, where # will be the frame number. Enjoy compositing!

5.5 Summary

Special Effects like morphing, dissolving, and compositing are very commonly used in movies and TV commercials. In this chapter we learned about the animation and image manipulation tricks used to achieve such effects. We saw how two digital images can be dissolved together to produce a resultant image with a compressed 256-color table. Warping allowed us to stretch, rotate, or squash images using control lines. The morph algorithm is really a combination of the warp and dissolve procedures. Morphing can be used to produce cool transformations between source and target images. At the end of the chapter we studied a simple method to composite two images into one combined image. The compositing procedure can produce some spectacular and surreal effects in animation.

You can use the software provided with this chapter to create your own special effects. The bitmap file format used by the programs is easily available and is standard for the personal computer platform.

CHAPTER 6

The Future

The previous few chapters have hopefully provided you with a strong foundation in the principles of three-dimensional computer graphics. We have provided adequate references and bibliography at the end of the book to give you an idea on where to start exploring more advanced topics that the book did not cover. In this chapter, we give you a glimpse of how computer graphics is affecting our lives today and what lies in the future.

The computer revolution has been the fastest revolution in the history of humankind. In just a decade, the computer has become a part of our lives, at home, at work, at recreational centers; in fact, we just cannot seem to avoid them no matter what we do. Over the last few years, computer graphics has emerged in its own right as an area of research and production. It has evolved from crude line drawings to beautifully realistic images rivaling scenes we can see in nature. Today, the science of computer graphics is concerned with more than pretty pictures. While research continues to advance the state of art in fields such as image processing, 3D modeling, animation, and rendering, there is also an active thrust to bring computer graphics into of our daily lives. We are indeed living in the computer graphics revolution age.

Where is computer graphics now and where is it headed? Let us look at some of the key areas where CG is being used and the areas where it is slowly gaining a stronger foothold.

Entertainment

The foremost area where CG is being used to its maximum is in entertainment. In the early 1980s, Lucas films started the trend with their trilogy *Star Wars*, for which computer scientists and artists skilled in 3D animation were used to generate much of the spectacular effects seen in these films. More recently in 1995, Pixar became the first production house to generate an entire movie using only computer graphics with *Toy Story*.

Computer graphics is now used in almost every Hollywood movie, be it the dinosaurs of *Jurassic Park*, or more subtle composite effects as seen in *Forest Gump*. The effects seen in films are going to become more widespread and more realistic as research in the area continues.

Game companies manufacturing CD ROMS and game cartridges rely on 3D graphics to achieve most of their effects. Even educational games try to illustrate concepts with the use of CG to make learning more fun.

Theme parks have started using computer graphics effects in a big way for their futuristic theme rides. These rides try to engulf the audience in a make-believe world that they are cruising through, defying laws of physics. The images displayed to the audience are typically computer generated and in perfect sync with the motion that they are experiencing, leading to a very *real* motion ride experience.

Virtual Reality

Virtual reality (or VR for short) is kind of a buzzword these days in computer graphics. VR is artificial reality created by a computer that is so enveloping that it is perceived by the mind as being truly real. VR exists in many forms. A traditional view of virtual reality uses headsets and data gloves. The headset serves as the eyes and ears to your virtual world, projecting sights and sounds generated by the computer. The data glove becomes your *hand*, enabling you to interact with this simulated world. As you move your head around, the computer will track your motion and display the right image. VR is the most demanding application for computer graphics there is, requiring hardware and software capable of supporting realtime 3D graphics.

At the moment most VR is being used to generate engulfing video games and interesting environments. The real interest of VR stems from the potential benefit to humanity. VR is being used to train pilots with the aid of flight simulators. VR will someday allow doctors to train on virtual *human bodies*, allow students to experiment with new ideas and concepts, and also let us explore strange new planets. People who cannot enjoy the real world due to physical handicaps can gain pleasure from enjoying simple pleasures like playing a golf game in a virtual world.

Medicine

Hospitals and medical research labs are relying more and more on CG to display images of internal organs and tissues. As technology progresses, it will soon be possible for doctors to view these organs as holographic images, as if the real thing were sitting in front of them. Medical imaging systems are all based on CG for their functioning. Research labs use CG to simulate molecular structures of DNA and other body building genes to explore the workings of our bodies and to experiment with them.

Defense

The Pentagon and the defense department spend a lot of money on 3D graphics research. They invest millions of dollars in developing flight simulators to train pilots to fly, simulating war scenes, and training personnel to fight and defend themselves. NASA too does extensive rocket simulations and testing on the computer before finally launching a real rocket.

Architecture and other Areas

Architects have started to create 3D walk-throughs of proposed buildings and architectures. This enables their prospective clients to see what they are getting before they finally give the green signal to the proposed design. In a few years, most architecture firms may create a VR world so the client can actually walk through his or her home even before it is built!

Combining CG with the mechanics of objects will allow companies to simulate and test the behavior of their products on the computer before they start expensive mass production. This idea is catching on with companies making all kinds of products, ranging from designing the shape of bottles for a shampoo to creating a detailed model of a car and testing its internal functionality.

So you can see that CG is gaining a foothold in all areas of our lives. As more and more research is conducted, the possible CG applications become more complex, and we shall continue to see this area grow and affect our lives. Indeed the future of computer graphics looks bright and promising.

Appendix A

Graphics File Formats

In this appendix we discuss some of the popular image file formats used.

CGM (Computer Graphics Metafile)

This file format was initiated and supported by the American National Standards Institute (ANSI). CGM file format was intended to be a device- and operating-system-independent format suitable for storage, retrieval, and interchange of pictures. It is widely used in CAD (computer aided design) systems. Its use is mandated by the CALS (Computer Assisted Logistics Support) program of the U.S. Department of Defense. The CGM format stores 2D raster image data with different color schemes. The data is organized sequentially in ASCII or binary and in general, no compression is used. CGM is rich in features (many graphical primitives and attributes) and it produces compact files. But the largeness of features makes implementation of CGM format quite difficult, though the usage of this format is spreading now.

EPS (Encapsulated PostScript)

This file format comes from Adobe Systems, Incorporated. The format is the description of a single picture encapsulated in another postscript file. The EPS description can also be read into another application specific program such as a paint program or document editor for combination with contents of that application. The EPS format is widely supported by several paint and CAD programs. The data stored is 2D raster geometry data and text and includes all the features of the postscript format including color specification, sequential data, and ASCII format.

FLI, FLC

This file format is used in the AutoDesk animator program. The format is used for efficient storage of frames to create animation. To create the animation, only differences between adjacent frames is stored. The data is stored as 2D raster data. The data is in binary format, sequential, and no compression is used. Color information is stored in the form of a color table.

GIF (Graphics Interchange Format)

This file format was initiated by CompuServe, Incorporated. The motivation to create this format was to efficiently transmit image data over telephone lines. This file format has widespread usage among CompuServe users and also among bulletinboard and newsgroup readers. The data is stored as 2D raster data in monochrome or 256 colors. The data is binary, sequential, and usually compressed using high performance data compression. Each pixel value in the image is stored and the data is stored in blocks of 256 bytes or less.

JPEG (Joint Photographic Experts Group)

The JPEG standard was adopted by the ISO (International Standards Organization) to create an international standard for digital encoding and compression of still images. The standard satisfies a broad range of applications. It has flexible compression ratios and has no restriction on the image contents. JPEG is a fairly new standard and its spread depends a lot on its acceptance in applications like fascimile, telex, medical, desktop publishing, graphics arts, and also scientific uses. The data format is usually 2D raster, sequential, binary with compression. JPEG is not a single format, rather a standard with a suite of 29 distinct coding processes.

MPEG (Moving Pictures Experts Group)

The MPEG standard was adopted by the ISO (International Standards Organization) to create an international standard to combine digital video data and audio data into a single sequential stream of data. The data is compressed to create acceptable audio/video performance. The format finds wide acceptance in compact discs, digital audio tapes, and many hard disks.

TIFF (Tag Image File Format)

This format was created with the support of many companies to provide for easy exchange of raster image data between application programs and also raster scanning devices. The design goals for the TIFF format was to support extendability, portability, and revisability. The format was common in desktop publishing and is now extending to video applications, fax transmissions, medical, and satellite imaging. It is now copyrighted with Aldus Corporation. The data is 2D raster format, random access, binary with support for black/white, gray, or color palettes. TIFF supports a large and diverse combination of image formats and resolution through the proper specification of image formats. Each data item in the TIFF file has a unique tag identifying its image type.

Appendix B

In this appendix, we discuss the model file format (mdl) which is used by the software accompanying this book can read.

All model files are in ASCII and start with the key word 'OBJFILE' to indicate that this is a file describing the objects in the scene.

Every object description starts with the key word 'STARTOBJ' and ends with the key word 'ENDOBJ'.

After the starting keyword, the object type is identified. We provide support for the following object types

- POLYG   polygon primitive
- NULLNODE   the Null object
- DISK   disk primitive
- CUBE   box primitive
- SPHERE   sphere primitive
- CYLINDER   cylinder primitive
- CONE   cone primitive
- GENERIC   generic objects
- GEOM   generic objects

Next comes the name of the object. After the name, the key variables defining the shape of the object are defined, as given below. Note that for some objects this is not applicable.

- POLYG   not applicable
- NULLNODE   not applicable
- DISK   radius
- CUBE   height, length, width
- SPHERE   radius

- CYLINDER   radius, height
- CONE   radius, height
- GENERIC/GEOM   not applicable

Next, for all objects, we define the default position as X,Y,Z coordinates, followed by the name of the parent object (if any) of this object.

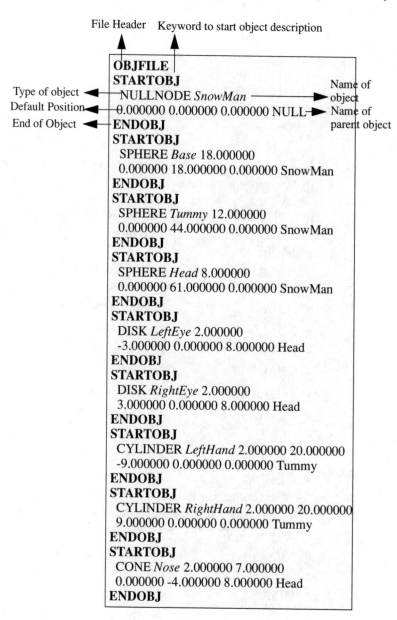

Fig. b.1: the snowman model file.

This name is set to NULL if the object is not a child of any other object in the scene. For objects other than the polygon and the generic object, we then end the object description by the keyword *ENDOBJ*. Fig. b.1 shows the file for the snowman model that we saw in the Modeling chapter, without the plateau object.

For polygons, we define the coordinates of its four vertices as X,Y,Z coordinates. A model file with a single polygon called Polygon_1 is shown below.

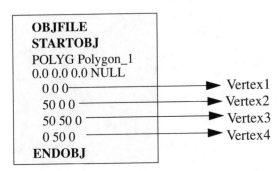

Fig. b.2: the polygon model.

Generic objects can be described in two ways (hence the two key words, *GENERIC* and *GEOM*). Remember that generic objects are nothing but a collection of polygons.

Those objects with the keyword *GENERIC* get defined as follows:

Each component polygon starts with the keyword *POLY* followed by the X,Y,Z coordinates of the four vertices, similar to the polygon object. Fig. b.3 shows a box shaped object with a length of 50 units, a height of 40, and a width of 30. We define such a shape as a generic object (although it can also be defined as a box primitive) in Fig. b.4, as a series of component polygons.

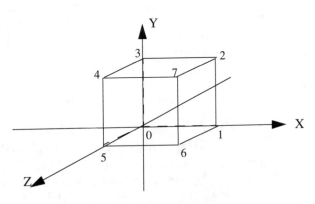

Fig. b.3: A box shaped object.

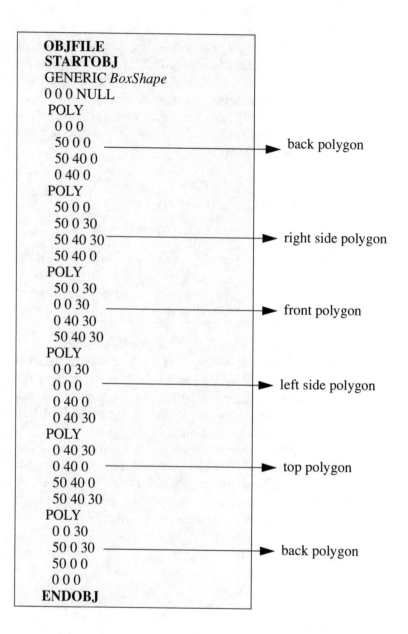

Fig. b.4: The generic object.

Those objects starting with the keyword *GEOM* are a bit more complex. First, all the vertices in the polygonal mesh are defined in a unique order. This is called the Vertex list. Then the polygons are defined by first giving the number of vertices making up the polygon (four in our case), followed by the positions of the vertices in the Vertex list that define the polygon. Referring to Fig. b.3, the 8 vertices marked 0 to 7 will appear in the vertex list. Fig. b.5 shows the model description with the keyword *GEOM* for such an object. It may be obvious from the figure itself that this representation of an object is more efficient in terms of file space. It is also more efficient to represent and deal with in the computer.

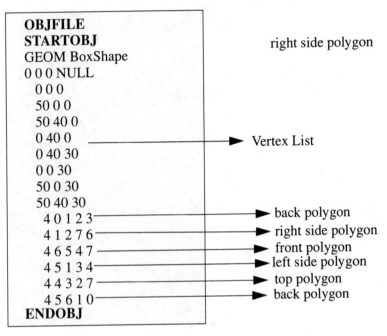

Fig. b.5: The generic object with the keyword *GEOM*.

This ends our discussion of the model file format that we used in this book.

Appendix C

Installing the Software

- Before proceeding to install the software, please close any other applications you may have running. This is to ensure that the Setup Wizard updates all the system files correctly.

- Insert the CDROM accompanying this book into your CDROM drive.

- If you are using Windows 95 (or Windows 98), go to the **Run** command under the **Start** menu.

- Type in the command **<drive>:\setup.exe** and press Enter (<drive> is the letter of the CDROM drive). For example, if the CDROM is in drive D:, type D:\setup.exe and press Enter.

Note: For other versions of Windows, check your manual on how to run Setup for an application.

- The 3dcg Setup Wizard will now appear. It will display an introductory message recommending you shut any applications you have running before proceeding. Assuming you have no other applications running, click on the OK button to begin the installation.

- You will be given the option to change the directory where the programs will be installed. By default, this is set to c:\3dcg. Do not change it unless you really have to, as this is the default directory we assume in the rest of the textbook. Click on the Computer icon to start installing the programs.

- Setup Wizard will check for necessary disk space. If there is enough space for the programs, installation will begin.

- Setup Wizard will inform you about the files it is installing on your hard drive.

- When it is done, you will get a message saying that setup was completed successfully. Click on the OK button to end the install.

- You will see a new folder under the Windows Explorer called **c:\3dcg** (assuming this is where you have installed your programs), which is where the executable programs reside. It is here that we store all the images and models, so it is useful to know how to get to this directory.

- Windows 95 and Windows 98 users will now see a new menu item called **3dcg** under the Start Programs menu. Under this group will be the four executable programs you will need for this book. To execute any of these programs, simply click on the appropriate option.

- Windows NT users should see a new program group created called **3dcg** which will have four icons for the four programs required. To execute any of these programs, simply double click on the appropriate icon.

Bibliography

What follows is a computer graphics bibliography. In addition to being a list of references for the contents of this book, it also contains references to books and journals, where the latest research in the field is being published.

Adams, L. *Visualization Graphics in C*, Mc-Graw Hill, 1991.

Anderson, S., *Morphing Magic*, Sams Publishing, Indianapolis, 1993.

Angell, I.O., *High Resolution Computer Graphics using C*, John Wiley & Sons, 1990.

Badler, N.I., B.A. Barksy, D. Zeltzer (ed), *Making Them Move: Mechanics, Control, and Animation of Articulated Figures*, Morgan Kaufman, Los Altos, CA, 1991.

Bertels, R., J. Beatty, and B. Barsky, *An Introduction to Splines for Use in Computer Graphics and Geometric Modeling*, Morgan Kaufmann, Los Altos, CA, 1987.

Bowermaster, J., *Animation How To CD*, Waite Group Press, Corte Madera, CA, 1994.

Cunnighamn, S., N. Knolle, C. hill, M. Fong, J. Brown, *Computer Graphics using Object-Oriented Programming*, John Wiley & Sons, 1992.

DeBoor, C., *A Practical Guide to Splines*, Applied Mathematical Sciences Volume 27, Springer-Verlag, New York, 1978.

Farin, G., *Curves and Surfaces for Computer Aided Geometric Design*, Academic Press, New York, 1989.

Farrell, J.A, *From Pixels to Animation*, Academic Press, London, 1994.

Foley, J., and A. van Dam, *Fundamentals of Interactive Computer Graphics*, Addison-Wesley, Reading, MA, 1982.

Foley, J., A. van Dam, S.K. Feiner, J.F. Hughes, *Computer Graphics: Principles and Practice*, Addison-Wesley, Reading, MA, 1987.

Glassner, A.S., *An Introduction to Ray Tracing*, Academic Press, London, 1989.

Glassner, A., *A User's Guide for Artists and Designers*, Design Books, 1994.

Gouraud, H., *Continuous Shading of Curved Surfaces*, IEEE Transactions on Computers, C-20(6), June 1971, pp. 623-629.

Heiny, L., *Advanced Graphics Programming Using C and C++*, John Wiley & Sons, 1993.

Joblove, G.H. and D. Greenberg, *Color Spaces for Computer Graphics*, SIGGRAPH 78, pp. 20-27

Lasseter, J., *Principles of Traditional Animation Applied to 3D Computer Animation*, SIGGRAPH 87, pp. 35-44.

Magnenat-Thalmann N., and D. Thalmann, *Computer Animation: Theory and Practice*, Springer-Verlag, Tokyo, 1985.

Morrison, M., *Magic of Computer Graphics*, Sams Publishing, Indianapolis, 1994

Newman, W.M., R.F. Sproull, *Principles of Interactive Computer Graphics*, McGraw-Hill, Japan 1981.

Porter T., and T. Duff, *Compositing Digital Images*, SIGGRAPH 84, pp. 253-259.

Prosise, J., *How Computer Graphics Work*, Ziff-Davis Press, Emeryville, CA, 1994.

Reeves, W.T., *Particle Systems- A Technique for Modeling a Class of Fuzzy Objects*, SIGGRAPH 93, pp. 359-376.

Sanchez J., M.P. Canton, *Computer Animation Programming Methods and Techniques*, McGraw-Hill, 1995.

Thomas, F., and O. Johnston, *Disney Animation, the Illusion of Live*, Abbeville Press, NY, and Walt Disney Productions, Burbank, CA, 1984.

Upstill, S., *The RenderMan Companion: A Programmer's Guide to Realistic Computer Graphics*, Addison-Wesley, Reading, MA, 1990.

Vince, J., *Three-Dimensional Computer Animation*, Addison-Wesley, Reading, MA, 1994

Index

abcissa 3
Adobe 110
algebra 12
algorithms
  fixed color palette 115
  median cut 115
  popularity cut off 112–115
  shading 95
  warping 119
aliasing 6
alpha blending 111
ambient reflection 87
  coefficent of (Ka) 87
AngY 92, 98
AngZ 91, 98
animate 51
animation 52
  camera 73
  Snowy 103
  traditional 52, 53
  viewpoint 71
animation.exe 58
anti-alias 6
anti-aliasing 6, 7, 15, 20
anticipation 69

back face 18, 25, 82
back face culling 25–26, 83
Back to the Future 41

backFace culling 40
Billy 123
bit 2–3
bitmap. See image file format,
    BMP.
BMP 98
bmp 110
bouncing ball 56
bounding box 34
box 21, 21–22
buffer. See video buffer.
byte 2

camera 75
camera match move 129
Camera View 15
camera viewpoint 93
cartesian coordinate system 12–
    14
  lefthanded 13
  righthanded 13
CD ROMS 132
clipping planes 40, 41
  back 33
  front 33
coefficients 88
collage 126
color map 127
color map. See color table.

color table 5
color-lookup table. See color map.
combination and specialized 100
complexity 23–24, 29, 82
complexity-to-performance 24
compositing 126
compression 8
cone 21, 28
Constructive Solid Geometry (CSG) 48
contour 47
control points 47, 48
co-planar 17
cubic curves 47
   bezeir 62
   b-spline 62
   hermite 62
cylinder 21, 27–28

diffuse reflection 87, 89, 90
   coefficient of (Kd) 87
digital image warping 118
digitizers 48
disk 21, 26–27
dissolve 116
dissolving 112
dithering 115
DNA 133
DrawDstLines 125
DrawSrcLines 125
dynamics 76

error range 127
extruding 47

field of view 33, 40
flat 99
flat shading 89
floating- point 3
Forest Gump 132
fps. See frames per second.
frame 54

frames per second 52
frequency table 113
front face 18, 25
frustrum. See viewing frustrum.

gallery 73, 102
generic model 29
geometry 12
graph 56
graphics images 7–8
GroundMat 102

hidden surface removal 82
hierarchical structures 42–43
human physiology 12

image
   background 128
   compositing 126–128
   dissolve 111
   enhancement 110
   foreground 128
   morphing 123–126
   warp 118
image file format
   BMP 7, 108
   EPS 7
   GIF 7
   JPEG 7
   TIF 7
image file formats 135
imitate 11
in-betweening. See tweening.
Internet 107
interpolation
   graph 56
   key frame 53
   linear 54–56
   non-linear 62
interpolation graph 58
Interview with a Vampire 123

Jurassic Park 132

key frame 52, 54, 57
kinematics 77–78
    forward 78
    inverse 78

light
    ambient 84
    distant 84
    point 84, 85
    SkyLight 104
    SnowLight 104
    spot 84, 85
    SunLight 104
local origin 21, 22, 26, 37
local origin. See origin.
Lucas films 131

mandrill.bmp 99
mantissa 3
material
    CarrotMat 103
    EyeMat 103
    SnowMat 103
    WoodMat 103
metamorphosing 123
Microsoft Corp. 108
model 11–12
model file format
    dxf 29
    mdl 29
model.exe 15
monochrome 4, 108
morphing 118
motion capture 76
motion ride simulators 72

NASA 118, 133
normal vector 18, 25
null node 43

optics 12
origin 12, 13, 16
orthogonal 18, 38
Paintbrush 8

palette 5
Pbrush 117
Pbrush. See Paintbrush.
Pentagon 133
pentagonal shape 12
persistence of vision 51
perspective 40
Peter 75
phi 22
photograph 11, 30
photorealistic 2, 81
PhotoShop 110
pinhole 31–35
pitching 36
Pixar 132
pixel 4, 5, 111, 120
pixels 4, 127
plateau 64
polygon 17
    multifaceted 18
polygon mesh 19, 23
popularity cut 112
postproduction 107–108
primitive 21
primitives 11
procedural 76
pyramid 33

RAM 3
realistic 91
rectilinear coordinate system.
            See cartesian
            coordiante system.
reflection
    ambient 87–89
    diffuse 87, 89–93
    specular 87, 93–95
rendering 1, 81
resolution 4, 5
revolution 47
RGB color model 3
RLE compression 108
Ross 123
rotation 36
rubber sheet analogy 118

rubber-banding 121

San Francisco 102, 126
scaling 38
scan conversion 6–7
scanners 48
schematic view 34, 36, 40
secondary action 71
shading
    Flat 95–96
    Gouraud 95, 96–97
single convex area 17
skateboard 75
smiley 118
SnowMan 43
Snowy 42, 64
solar system 11
source image 116
specular 93
specular reflection 87, 93
    coefficient of (Ks) 87
sphere 21, 22–26
squash and stretch 67
staging 68–69
staircase effect 6
Star Wars 132
surfaces of revolution 47

target image 116
texture mapping 98–100
    combination and specialized
        99
    flat 99
    tiling 99
    wrapping 99
theta 22
tiling 102
timing 70
Tom Cruise 123
Toy Story 1, 82, 132

tradeoff 19, 24
transformation
    matrix 39, 41, 72
transformation matrix 56
transformations 35–41
    camera 40
    object 35
transition 111
translation 35
transparent
    pixels 127
tree structure 42
triplet 3, 83, 86
true-color system 111
tweening 52

vertex 12, 17
VGA 5
video buffer 4, 5
view volume 33
viewfinder 2
viewing frustrum 33, 34
virtual reality (VR) 132
visual realism 81

walk-throughs 133
WallMat 102
warping. See digital image
        warping.
wire frame 20, 81
world coordinate system (WCS)
        14
world origin 21, 22, 26, 37
world origin. See origin.
World Wide Web 1
wrapping 100

yawing 37

z-buffering 83